AWESOME ROBOTICS PROJECTS
for Kids

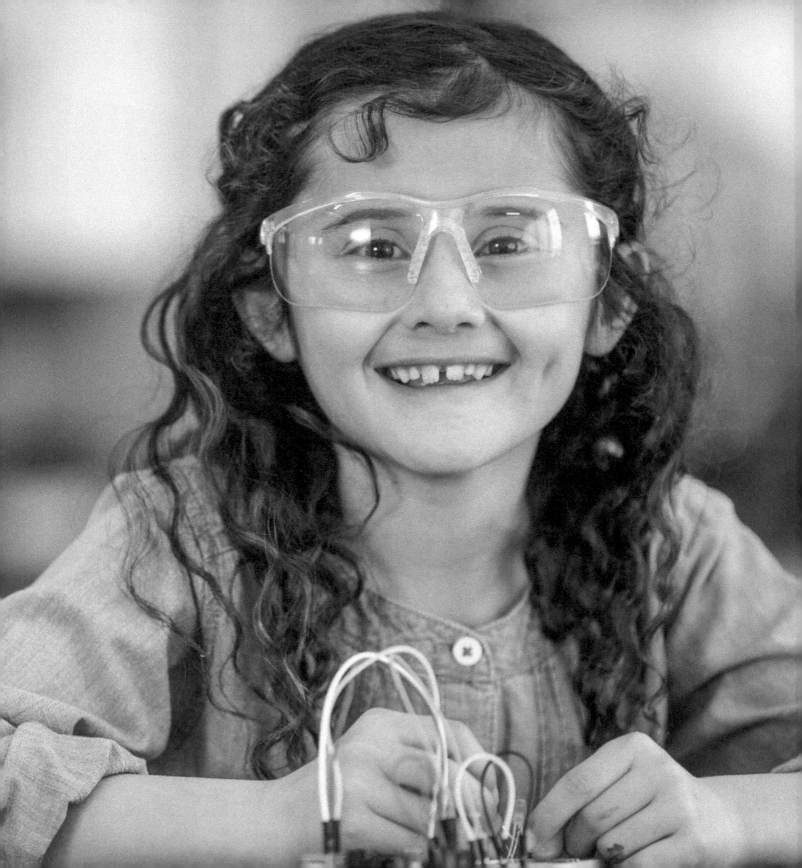

Awesome ROBOTICS PROJECTS for Kids

20 Original STEAM Robots and Circuits to Design and Build

BOB KATOVICH

ROCKRIDGE PRESS

For general information on our other products and services or to obtain technical support, please contact our Customer Care Department within the U.S. at (866) 744-2665, or outside the U.S. at (510) 253-0500.

Rockridge Press publishes its books in a variety of electronic and print formats. Some content that appears in print may not be available in electronic books, and vice versa.

TRADEMARKS: Rockridge Press and the Rockridge Press logo are trademarks or registered trademarks of Callisto Media Inc. and/or its affiliates, in the United States and other countries, and may not be used without written permission. All other trademarks are the property of their respective owners. Rockridge Press is not associated with any product or vendor mentioned in this book.

Interior and Cover Designer: Patricia Fabricant
Photo Art Director/Art Manager: Sara Feinstein
Editor: Orli Zuravicky
Production Editor: Chris Gage
Photography © 2019 Bob Katovich, cover, p. 19, 21–35, 39–53, 57–73, 75–85, 89–105, 109–129; iStock, p. ii, viii, x, 8, 54, 86, 106, 130; Shutterstock, p. 16, 36.

Author photo courtesy of © Nicholas Souder.

ISBN: Print 978-1-64152-676-0 | eBook 978-1-64152-677-7

R0

FOR KENT

CONTENTS

So You Think Robots Are Pretty Cool, Huh?

Me too! But have you ever asked yourself *why* robots are cool? You must have wondered once or twice, because now you own this book! I'm about to tell you exactly why robots are so awesome. My name is Bob Katovich, but you can call me Bob Kat. I am the program director of a robot store in Chicago called Robot City Workshop. I set up robotics workshops for kids of all ages, and I invent robots that are super fun to build.

This book is going to help you make your own robots. There are 20 robot projects for you to do. They start off pretty simple and get more complex as the book goes along. If this is your first time building robots, I suggest starting at the beginning and going through the projects in order. But if you already have some experience with robotics, feel free to skip around. The projects are broken into different categories: robots at home, robots for entertainment, robots in space, robots on the job, and robots in the operating room. Each project relates in some way to the chapter in which you'll find it. Some of the projects use the same parts, so you can take apart one robot and use the pieces to build another one. Other projects build on the previous projects, so at the end of a chapter, you will build a cool robot using what you learned in that chapter—or even in an earlier one.

I hope that you enjoy this book and that it inspires you to continue your robotics education!

—BOB KAT

Chapter 1

WHAT IS ROBOTICS, ANYWAY?

If you are reading this book, you probably think robots are cool. And you are right. Robots *are* cool. But what do robots do? What makes a robot cool? Why do we need robots? Before we answer those questions, we should think about humans. Why are humans cool? What is special about humans that robots cannot do? For one, humans can love one another. Humans can also laugh with one another. Humans can appreciate and enjoy life. But sometimes, humans can get frustrated or mad. To enjoy life, we often have to do things that are difficult or boring, like work. Sometimes, work can be dangerous. That's where robots come in. They can help us perform difficult, boring, or dangerous tasks. Then, we humans can focus on what's really important: loving one another, enjoying life, and laughing at Internet memes.

Studying **robotics** can help us continue to grow as a **civilization**. Thousands of years ago, the ancient Egyptians built the pyramids. They remained the tallest buildings for thousands of years. Also, they were made using human labor. The pyramids took many, many years to complete and cost thousands of lives. Today, we use science, technology, engineering, art, and mathematics, or **STEAM** for short, to build skyscrapers. The advances we've made in these areas help us build better things in less time and much more safely. With the use of STEAM, we can send humans to the moon, and get them back to Earth. We can also build awesome computers and game machines, like Nintendo consoles.

Did you know that robots help us stay alive? They help grow food to eat. Robots help take care of us when we are sick. Robots help people who cannot help themselves. Stephen Hawking, a famous genius, was paralyzed. He used a robotic wheelchair to move around, and he spoke through a computer using a robotic voice. With the use of these robots, he helped us better understand the universe.

In this book, we are going to learn about what makes a robot a robot, and we'll learn all about how to build our own. Are you ready to build some robots?

ROBOTICS THROUGH THE AGES

Compared with the age of the universe, or the number of years humans have been around, robots are a pretty recent development. But people have been dreaming about robots for thousands of years. Before electricity was discovered, early "robots" were called **automatons**. These were self-operating machines that used **gears** and **levers** to move in planned ways. Ancient Greeks had a myth about a giant bronze automaton, named Talos, that protected the island of Crete from pirates. (Now there is a ride at Disneyland, Pirates of the Caribbean, full of **mechanical** automaton pirates.)

A bit more recently than the time of ancient Greece, Leonardo da Vinci, the famous **Renaissance**-era painter and inventor, created plans for a mechanical knight **designed** to move its head, raise its arm, and sit up. But early automatons simply could not compete with human abilities. Automatons became popular during the Renaissance as entertainment, mostly for royalty. A famous builder of autom-atons, Jacques de Vaucanson, created the Digesting Duck in 1739. This mechanical duck could eat kernels of grain, which it appeared to then process and discharge out the other end. Now that's a messy way to change a robot's batteries!

A 1920 play by Czech writer Karel Capek, called *R.U.R.: Rossum's Universal Robots*, intro-duced the word *robot* to the world. The "robots" were mechanical servants that looked and acted just like we do. The play did not end well for the humans.

The first true robot was Shakey, developed at Stanford University during the 1960s. It could roam around, using cameras and **bump sensors**, just like the Roomba vacuum robot. It was called "the first electronic person" by *Life* magazine in 1970. While Shakey wobbled around, companies everywhere were begin-ning to use robotic arms to transform how they made products.

The **Space Age** brought us many robots. In 1959, the Soviet Union sent a robotic **space probe**, Luna 3, to photograph the dark side of the moon for the first time. It also landed the first robotic lunar rover in 1970. In 1977, the United States launched Voyager 1 and 2, the longest-running, fastest, and fastest-traveling space robots to date. Both are now billions of miles from Earth.

In the 1980s, the idea of a robot butler became reality with the Omnibot 2000. Built by the Japanese company Tomy, Omnibot was a futuristic-looking robot that could carry small objects and play cassette tapes. Unfortunately, if you lost its remote control, you would have to wait until the late 1990s, when you could buy a used one on eBay. Popular science fiction, such as *Star Wars*, also led to the development of robots designed for entertainment and learn-ing, such as the HERO—short for Heathkit Educational Robot—series. In the mid-1990s, the car manufacturer Honda developed P3, a **humanoid** robot that could walk, wave hello, and shake hands.

The 1990s saw the release of Furby, a small, furry, language-learning toy robot. It started off speaking its own weird language when first taken out of its box, but it would begin saying English words over time. Furbies were so in demand that stores often ran out before they could restock the shelves. While toy robots became increasingly popular, robots were also developed to do everything from explore rough terrains, including that of Mars, to study how fish swim. A robot head named Kismet could even display emotions.

In 2000, the da Vinci Surgical System revolutionized the medical field. It had four robotic arms—three held surgical tools and one had cameras—that surgeons could use to perform complex surgery by making only tiny incisions. That same year, the Sony Dream Robot was created, with the ability to recog-nize different faces, express emotion, and walk.

The following year, another robot was created to help assemble the **International Space Station**.

The field of robotics will continue to grow and advance because there are so many areas of life in which robots can help humans do better, smarter work. With all areas of STEAM working together, the future of robots is wide open.

SO . . . WHAT'S A ROBOT?

If you're still wondering what a robot is, that's okay. You might be thinking, "Isn't a robot just a machine?" That's okay, too. Let's look at the differences between a robot and a machine. A machine performs a simple task that is automated, or repeated. It might even be able to perform many different tasks. But it always does the tasks the same way, and it performs them until someone tells the machine to stop or turns it off. A robot also performs tasks, but a robot can react to its environment and change what it does based on certain factors. A robot can "sense" its surroundings, then "think" about how to "act" on that physical information, or **input**.

A robot performs these actions using different parts. Let's talk about the parts of a robot and find out what they do.

SENSOR

A **sensor** takes input, or physical information, from its environment and turns it into an **output**, usually in the form of an **electric signal**. There are many different types of sensors. Many are similar to our own senses. **Photoresistors**, **infrared LEDs**, and **solar panels** detect light, just like our eyes do. **Microphones** detect sound, just like our ears do. **Barometers** detect changes in pressure, and **thermometers** detect changes in temperature, like our skin does. In the same way that our noses can detect smoke and some gases, detectors can sense different gases. Some robots can detect gases that even our noses can't smell.

CONTROLLER

The output from the sensor becomes the input for the **controller**. These input signals from the sensor are **decoded** by the robot's controller. Decoding is a type of "thinking" that is based on how a controller is **programmed**, or instructed. A **microcontroller** can convert, or turn, a simple input into a simple output. A **computer** can perform much more complicated conversions. In humans, our brains act as our controllers.

ACTUATOR

The output from the controller becomes the input to move the **actuators**, or the **devices** that allow certain parts of a robot to move. They use levers and gears, **motors** and **servos**, **hydraulics** and **pneumatics**, or **solenoids**. In humans, our muscles are our actuators.

EFFECTOR

The actuator moves the **effector** that interacts with the environment. Arms, legs, and fingers are both human and robotic effectors.

POWER SOURCE

Robots need a source of power. **Batteries** are the most common robotic power source. A battery produces an electric current. This current travels to the different robot parts in a path called an **electric circuit**. Some robots are powered by the sun, through *solar power*. Many space robots are powered by a **nuclear reaction** that produces heat. This heat is turned into electricity by a device called a **Peltier junction**. Human digestive systems create heat and electrical energy, too, when we eat.

Understanding the various parts of a robot is essential if we're going to build our very own robots. Now we understand that a robot is a mechanical object that can sense its surrounding and determine and carry out simple or complex actions based on its design and capabilities. Like humans, robots need a power source, senses, a brain of sorts, and parts to carry out movement.

WHAT ISN'T A ROBOT?

Q: What's the difference between a toaster and a Roomba?

A: Toasters burn your bread, but you have to burn a lot of bread to buy a Roomba! HEY-O!

(If you didn't understand that joke, "to burn a lot of bread" means to spend a lot of money.)

Many people debate or disagree about what is and is not a robot. Is a toaster a robot? You set its dial (Is that a sensor?), then press a lever (Is that the input?) that turns on the heating coils (Are these actuators?). After a certain amount of time (Is that a controller?), it pops the bread out (Is that an effector?). Many **roboticists** see a toaster as nothing more than a simple machine. But what is a robot, if not just a complex machine? As with many definitions, it depends on the human. Someone who helped build a fancy space robot might be offended to hear someone call an automatic flushing toilet a robot. But someone who has never seen a toilet flush automatically might bow in respect and say, "Thank you, poop-disappearing robot!"

Some people differentiate a machine from a robot by figuring out whether the machine or robot is capable of responses that are specific to the input it receives. Machines can do only what they are programmed to do, how they are programmed to do it. They do this every time someone pushes a button or enters a command, with no change in response (unless, of course, they are not working correctly). The action is automatic. A robot, on the other hand, responds to **stimuli** more like a human might. We assess the situation, determine the best response, then use our brain to control our movement in response. Likewise, the argument is that robots engage in some level of assessment (or thought) based on the inputs—or environmental factors—to generate an appropriate output, or action. Your microwave can heat your restaurant leftovers for however many minutes you tell it to; it can't, on the other hand, sense that your leftovers are still warm, or how full the container is, to come up with an ideal number of minutes for which to reheat your food on its own. This makes it a machine, not a robot.

One important thing to remember is that all robots start as machines. The addition of sensors and programming, which allows a robot to use those sensors to determine how to act, is what makes it *more* than a simple machine. Many of the "robots" you will build in this book are simple. This will help you develop the building blocks to one day, perhaps, create your own high-tech robot to serve as your personal assistant, take you to space, or maybe just clean your room!

WHAT'S THE DEAL WITH ROBOTICS AND STEAM?

Robotics is an important subject because it uses every form of STEAM education. But what is STEAM? STEAM is an **acronym**—a word in which each letter stands for another word, like YOLO (you only live once) or BRB (be right back). As I mentioned earlier, STEAM stands for **science**, **technology**, **engineering**, **art**, and **mathematics**—all subjects that are needed to design and make robots. And because robots not only make awesome toys but also have the ability to improve our lives and our world, robotics is now being taught in schools.

Let's take a look at how robotics and STEAM connect. **Science** uses observation and experiments to learn about the world around us. This can lead to **innovations**, or breakthroughs, in robotics. (Or people can build robots to perform their experiments!) **Technology** means "science of craft." To perform experiments, scientists have to craft, or build, their own tools, sensors, detectors, and controllers. Many of the robot parts we use today came from past science experiments. Only afterward did people realize how useful they could be. **Engineering** uses rules learned in science to build everything around us, including our robots. **Art** isn't just about how things look; art helps people think creatively. It helps train your brain to think about a problem from all angles, and to find different ways to make things fit and work together. **Mathematics** is the language scientists use to describe how the universe works. Any measurement or quantity uses mathematics. Robots must use every STEAM discipline together.

During the early US space program, the computer hardware company IBM delivered a state-of-the-art computer (at the time) to help **NASA**—the National Aeronautics and Space Administration—calculate the very complicated mathematics needed to get **satellites** and astronauts into orbit. The computer took up an entire room, but because someone forgot to use mathematics to calculate the size of the door into the room, the computer wouldn't fit! In robotics, every part of STEAM must work together; otherwise, like NASA, you may find yourself tearing holes in walls to make your inventions fit!

Chapter 2

HOW TO USE THIS BOOK

This book doesn't just teach you about robotics. It is also a do-it-yourself guide to building cool robots! Many of the robots build on each other, and many use the same parts, like light bulb circuits, motors, and solar panels. If you choose to, you can use some parts from earlier robots to build later robots, or you can build each robot individually. It's entirely up to you (and your adult helper).

The projects start very simply. First, you'll learn how to build your own circuits, because you have to know how a robot is powered before you can build one. As the book progresses, the projects will get more complicated. Once you learn how to build circuits, the more advanced robot projects will show you how to build things that move, first with vibration motors and then with wheels and even fans. There is also a project that doesn't move at all—it just makes weird robot noises. (How awesome is that?!) Toward the end, you'll learn about walking robots, building your own four- and six-legged robots. Finally, you will make a robot that walks on two legs, just like you.

I told you robots were cool!

GETTING READY

Now that you're ready to build robots, where do you start? First, you need a clean, flat surface, like a table or desk. Building on the floor may work for some projects, but you don't want to get all bent out of shape scrunching over for a long time. Some of these projects will take you more than an hour to complete. You want to be sitting comfortably and upright. A table or desk also lets you spread the parts out without anything getting lost in the carpet or eaten by the dog. (Seriously, keep all parts away from pets.)

Which project should you build first? You can start at the first project and go through each one in order. If this is your first time building a robot or circuit, I suggest building the robots in order. If you have already built a circuit, feel free to skip those projects and move on to something you haven't built before. You can jump around, pick robots you think are cool, and build those.

If you jump ahead and find a robot that is too challenging, that's okay. It is okay to fail. Most of these robots didn't work at first. Many of them had to be built a few times to get them working the right way. In fact, it is rare to build a robot that works after the first try. A big part of building robots is **troubleshooting**, or figuring out what is wrong when something *doesn't* work, and then fixing it.

For each project, there is a list of materials and tools. You will also find any cautions you should be aware of and keep in mind before beginning to build. Some projects require adult help or supervision. Some use sharp tools or power tools, like a drill. Some use hot glue or super glue. Be careful. *Always* ask an adult for help. Some people actually say that knowing when to ask for help is a true sign of intelligence. So go ahead, be smart. And remember: Any complex robot you have seen out in the world had a team of people building it—not just one person. This can be a great opportunity to get your friends involved. Learning to work as part of a team is necessary for a future in robotics, so the sooner you learn to work with others, the better off you'll be.

You don't need to dress in a jumpsuit like a NASA engineer to build robots, but if it makes you feel good, do it. When it comes to choosing what kind of clothes to wear when building robots, I have only one hard-and-fast rule: Wear some. Having said that, the right clothing will help you, so here are some guidelines I think you should follow. Avoid baggy clothes, which can get caught on parts, or on your half-built robot, sending it flying off the table. (I have done this before. Lesson learned.) Always wear shoes. You don't want anything heavy or sharp falling onto your toes. Safety glasses are recommended if you are cutting anything or using power tools. Plus, they make you look like a legit roboticist.

Gather all the parts and tools you'll need for each project before starting. It can be frustrating to get halfway done, or even almost entirely done, with your project only to realize you are missing one piece. Checking your materials before you start can help you avoid this common frustration. It's also easy to get distracted if you have to get up and look for a tool or part in the middle of a robot build, which could lead to mistakes down the line. Finally, be sure to read through the instructions for each project before starting so you are ready for each step.

DOING THE PROJECTS

A very important part of building robots (and furniture) is learning how to follow instructions. More so, learning how to turn an idea or drawing on a flat piece of paper into a three-dimensional, physical thing is **empowering**. It makes you feel powerful and confident, like a science wizard.

Each project includes step-by-step instructions, as well as pictures. While each project has many pictures to help guide you, you will notice that there aren't pictures for every step of every project. Also, keep in mind that because one numbered step can include several different smaller steps within it, some photos show only one or two parts of a step rather than the whole step. If you are confused by a step, that's okay. You can ask an adult for help. (Just make sure they don't take over building the robot for you.)

Here's a trick that may come in handy: When looking at the picture of a step, place your parts down on the table in the same order and orientation as they are shown in the picture. Make sure the parts in the picture and your parts look the same.

Each project has a "How it works" section. Learning how the robot works can help you troubleshoot your robot if it is not working. It can also show you why your robot does what it does.

You will also find a "STEAM Connection" section for every project. This helps show how what you're doing is using and building upon your STEAM knowledge. If you are testing something, like different circuits, where to put your solar panel, or what surface works best for your robot to move, you are being a scientist. Building circuits or electronics and using power tools, are forms of technology. If you are connecting parts together or troubleshooting, you are engineering. Designing and decorating your robot is art. And measuring with a ruler and counting parts requires mathematics. Knowing what step of the robot-building process uses each subject reminds us that all of these fields need to work together for our world to continue to improve through robotics.

SAFETY

Each project has a "Caution" section. This is one of the most important parts of robot building. If someone tells you about a way to be safe, usually it's because they hurt themselves doing the opposite. Learn from other people's mistakes, and stay safe. Some projects tell you to get adult help for steps that require sharp or power tools. Always ask an adult for help with these steps, even if they tell bad jokes. If you find yourself frustrated, seeking help is also a good idea. It is very easy to get hurt when you're fuming because something isn't going as expected, and sometimes a new set of eyes can help you tackle a problem you're having trouble seeing.

Enhance the safety of your work environment by starting with a clean, dry, uncluttered surface.

Avoid having drinks, or anything that could spill, nearby. That includes your fish bowl, if you have one. A surge protector and tools with rubberized handles are also good ideas when working with electronics. Being organized can also help keep you safe. If you have a safe place for everything, there's less likelihood of, say, a tool getting budged off your desk and falling on your foot or your cat! (Seriously, keep your pets out of your robotics workshop.) Enhance your own safety by making sure you're appropriately dressed, with fitted clothing, shoes (preferably with rubber soles), and safety goggles.

WHERE TO FIND PARTS

Many of the robot parts used in these projects, like batteries, wire, small screws, and magnets, can be found around the house. Others can be found at local hardware stores or other local shops. Don't be afraid to get creative. Ask a local pizza shop for a pizza saver tripod for the scribbler bot, or ask your doctor for some tongue depressors. Some parts, like motors, solar panels, or battery holders, are easier to buy online. Following is a list of great places to find robot parts for the projects we'll be doing.

- **American Science & Surplus** (www.sciplus.com) is a wonderland for all sorts of weird stuff. Stores in Chicago and Milwaukee have aisles of motors, electronics, materials, containers, and other odd robot parts. Many of the parts in these projects—like the small bulbs, **bulb receptacles**, containers, and alligator clips used in Chapter 3—can be found there or ordered from the website.

- **Adafruit Industries** (www.adafruit.com) is an online marketplace where you can find all sorts of electronic components, motors, batteries, microcontrollers, battery holders, and solar panels. It also offers **tutorials**, or guides, and books about building cool electronic projects for kids.

- **Microcenter** (www.microcenter.com) is a retail and online store that sells computers and computer parts. But it also has a great "DIY/Maker" section (under "Products," then "Electronics"), where you can buy many parts for the projects in this book.

- **Pololu** (www.pololu.com) is a robotics and electronics dealer. Many hard-to-find robotics parts, like gear motors, wheels, and **ball casters**, can be ordered here.

- **Robot City Workshop** (www.robotcity workshop.com) is a place to learn about robots and build them in a workshop. It also has a retail and online store that sells robot kits and robot parts.

- **Amazon** (www.amazon.com) and **eBay** (www .ebay.com) are great places to buy materials, like tongue depressors, wooden skewers, O-rings, and screws.

SUPPLIES LIST

The supplies list has been broken down into categories to help you as you go about finding the items you'll need to make the robots in this book. This is a complete supply list, but note that many of the parts for one robot can be reused to make other robots. If you plan to make and keep all your robot projects, this is a good shopping list. If, on the other hand, you think you might want to reuse some of your supplies from one project to the next, it might be a better idea to take it project by project, following the project-specific materials lists instead of going out and getting everything on this list. There are also some supplies on this list that may be very difficult to find in stores, but if you keep your eyes open as you go about your days—say, when going out for pizza with your family—you'll find everything you need to build your robots soon enough.

TOOLS & BASICS

- ❏ Drill with two different-size drill bits
- ❏ Hammer
- ❏ Hobby knife
- ❏ Hot glue gun and glue sticks
- ❏ Ruler
- ❏ Scissors
- ❏ Regular-size screwdriver
- ❏ Small screwdriver
- ❏ Table or desk
- ❏ Wire stripping tool or wire cutter

ARTS & CRAFTS SUPPLIES

- ❏ 2 alligator clips
- ❏ 1 or 2 rolls of aluminum foil tape
- ❏ Cellophane (transparent) tape
- ❏ 17 (6-inch) craft sticks
- ❏ 6 (1½-inch) pieces of craft sticks
- ❏ Double-sided cellophane (transparent) tape

- ❏ Double-sided foam tape
- ❏ 1 piece of fabric, a couple of inches wide and as long as the **circumference** of a bowl
- ❏ 4 glue dots
- ❏ 5 googly eyes (or more!)
- ❏ Regular-size marker
- ❏ 3 small markers
- ❏ Paint, stickers, or other decorations (optional)
- ❏ Pencil
- ❏ 56 (6-inch) tongue depressors
- ❏ Vinyl electrical tape
- ❏ Wooden skewers or dowels, ⅛ to 3⁄16 inch in diameter (for making a total of 21 rods ranging in length from ¾ to 4 inches)

BATTERIES & BATTERY COMPONENTS

- ❏ 21 AA batteries
- ❏ 2 AAA batteries
- ❏ 3 (9-**volt**) batteries
- ❏ 6 (2-AA) battery holders with on/off switch and wire leads attached
- ❏ 4 (2-AA) battery holders with on/off switch
- ❏ 1 (2-AAA) battery holder with on/off switch
- ❏ 3 (9-volt) battery snap connectors
- ❏ 3 CR2032 3-volt button cell batteries
- ❏ 5 small solar panels, rated 0.5V, 800mA, or higher

ELECTRICAL COMPONENTS

- ❏ 2 small, 170-point **breadboards**
- ❏ (1-**IF**) capacitor
- ❏ 1 (100-μF) **capacitor**
- ❏ 3 Elenco 2-in-1 **gearbox** kits
- ❏ 1 high-speed gearbox
- ❏ 1 low-speed gearbox
- ❏ 1 photoresistor
- ❏ 3 (150-**ohm**) resistors
- ❏ 1 (470-ohm) **resistor**
- ❏ 1 (1,000-ohm) resistor
- ❏ 1 (1 million-ohm) resistor
- ❏ 1 momentary push-button switch
- ❏ 1 small speaker with wires attached
- ❏ 1 555 timer integrated circuit

FOUND OBJECTS

- ❏ 1 plastic bottle (with dimple in its cap)
- ❏ 1 counterweight (such as a small padlock)
- ❏ 1 cardboard tube
- ❏ 1 square piece of cardboard, just bigger than a bowl
- ❏ 1 small Chinese takeout container
- ❏ 1 small circle of thick paper (like from a paper plate) covered in aluminum foil
- ❏ 1 small domed lid
- ❏ 1 (16-ounce) plastic jar, emptied and cleaned
- ❏ 1 small paper or Styrofoam bowl
- ❏ 2 pizza saver tripods
- ❏ 2 rubber bands
- ❏ 2 to 4 plastic sauce containers (from local restaurant)
- ❏ 2 springs
- ❏ 1 small square tin with lid
- ❏ 2 small squares of Styrofoam
- ❏ 1 wire clothes hanger

HARDWARE STORE SUPPLIES

- ❏ 2 adhesive-backed foam washers
- ❏ Adhesive-backed magnetic tape
- ❏ 1 ball bearing or other small, round object
- ❏ 4 plastic ball casters
- ❏ 4 (¼-inch by 3-inch) bolts
- ❏ 2 small bolts with washers and nuts
- ❏ 1 box of small screws
- ❏ 1 box of small washers
- ❏ 12 small **cotter pins**
- ❏ 3 medium rubber **gaskets** for tires (that fit on small wheels)
- ❏ 64 small rubber O-ring gaskets, ⅛ inch in inner **diameter** (that fit snugly on wooden rods)
- ❏ 1 small machine screw with washer and nut
- ❏ 1 small metal base
- ❏ 2 small **neodymium** rare earth magnets
- ❏ 1 lock nut (that fits ¼-inch by 3-inch bolt)
- ❏ 7 regular nuts (that fit ¼-inch by 3-inch bolt)
- ❏ 2 small nuts or lock nuts
- ❏ 4 ring connectors

LIGHTS

- ❏ 7 mini 1.5-volt bulbs
- ❏ 7 mini plastic bulb receptacles
- ❏ 4 LEDs

MOTORS

- ❏ 3 dual shaft gear motors
- ❏ 1 **low-voltage**, **low-current** DC motor (with wires)
- ❏ 2 vibrating disk motors
- ❏ 1 large vibration motor

TOY PARTS

- ❏ 1 toy boat
- ❏ 2 **propellers** (that fit on the low-voltage, low-current motors)
- ❏ 1 small robot or astronaut figurine (about 1½ inches tall, such as a LEGO minifigure)
- ❏ 3 wheels (best if 2 fit the low-speed gearbox and 1 fits the high-speed gearbox)
- ❏ 2 wheels (that fit on the dual shaft gear motor's oval shafts)
- ❏ 2 small plastic wheels with spokes
- ❏ 2 thin plastic wheels

WIRES

- ❏ 12 jumper wires (any short length of solid core copper wire)
- ❏ 2 (6-inch) pieces of solid core copper wire
- ❏ 1 (6-inch) piece of black wire, stripped on both ends
- ❏ 1 (6-inch) piece of red wire, stripped on both ends
- ❏ Solid core copper wire

Chapter 3

ROBOTS AT HOME

When kids are asked, "If you could have any robot to help you in your daily life, what would it be?" one of the most popular answers is: "A robot to do my homework." Not to burst your bubble, but you will never become a robot genius if you have a robot do all the learning for you. Part of the reason people get into robotics is because it challenges us to think outside the box and solve tough problems. The second most popular answer to the question, "What robot would you want?" is: "A robot to clean my room." Now *that* we can do.

In fact, history is filled with geniuses who have people to clean up after them while they're busy focusing on being geniuses. Who has time to consider the mysteries of black holes *and* clean up the black hole that is a messy bedroom? But people must be paid (or in the case of siblings, maybe bribed). Fortunately for us, robotics is constantly making advances that help make our home lives easier. Think of the Roomba, the disk-shaped vacuum cleaner that uses intelligent sensors to vacuum a room on its own, or the LawnBott, which will have your lawn mowing duties done in no time.

Robots for home use can do much more than clean. Some can serve as your personal assistant, recommend and play entertainment based on your preferences, and even offer companionship. There is even a robot, called the Jibo, that can help facilitate communication between members of a household. This little WALL-E look-alike can learn from everyone with whom it interacts and remember what they say. So, if you were too busy finishing up your homework over breakfast to remember what your mom said about plans for this coming weekend, Jibo has your back.

While high-tech robots have become a common part of our lives, robots that help us out in our daily lives are perhaps the most impactful, freeing up time that we might otherwise spend doing unappealing chores or time-consuming tasks, so that we can engage in more meaningful and enjoyable activities. Some tasks, of course, are more unsavory than others. But whether you're cleaning up your room, mowing the grass, or simply turning on a light, all of these little and not-so-little tasks can add up to a big chunk of your free time. With the help of robots, you can reclaim that time so you can, say, build more robots.

In the following projects, we will learn the basics of how robots are powered. Then, at the end, we will use that knowledge (and those parts) to build a cool robot night light for any room in your home.

BUILD A SIMPLE CIRCUIT

Before we build a robot, we need to know how a robot is powered. Most robots run on **electricity**. The electricity travels to the robot's motors, lights, or brain through an electric circuit. But what is a circuit? Think of a shape that sounds (and looks) like the word *circuit*. If you thought of a circle, you are on the right track. A circuit is like a circle, or loop, of electricity. Any battery has two **terminals**, or points to connect. They are usually marked with a plus (+) or minus (–) sign and are labeled positive and negative. To power anything, what you are connecting must be connected to both sides, or terminals, of the battery. In this project, we are going to build a super-simple circuit and make a light bulb glow.

TOTAL TIME: 5 MINUTES

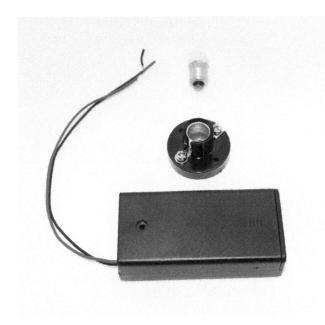

MATERIALS:

- 1 (2-AA) battery holder with on/off switch and wire leads attached
- 2 AA batteries
- 1 mini plastic bulb receptacle
- 1 mini 1.5-volt bulb
- Small screwdriver
- Wire stripping tool or wire cutter

CAUTION: You may or may not need to use a wire cutter or wire stripping tool for this project. It all depends on whether enough length of wire is exposed for you to work with. If you need more length of wire, ask an adult for help. Wire cutters/stripping tools can be very sharp.

CONTINUED ➡

STEPS:

1. Make sure there is about ½ inch of metal wire exposed on the battery holder wires. If you need to strip, or expose, more wire, have an adult help you, as wire cutters/stripping tools can be sharp. Wires have two parts: the **conductor**, which is metal, and the **insulator**, which is the plastic coating around the metal wire. The conductor carries the electric current. The insulator protects the conductor, so you can touch the wires without worrying about getting shocked. You should not get shocked on any of these projects because the **voltage** and current are very low. Still, wires are made this way to protect us, just in case. Insulators come in many colors to help us remember which wire is doing what part. For batteries, usually the red wire is positive (+) and the black wire is negative (–).

2. Slightly unscrew both screws on the mini plastic bulb receptacle, or holder. Look at the screw terminals on the bulb receptacle. On the one we use for this project, one terminal is square, or positive (+), and the other is round, or negative (–). If your bulb receptacle is different, the negative terminal is connected to the sides where the bulb screws in. The positive terminal is connected to the contact where the bottom of the bulb touches the center. Twist the metal from the red wire around the screw above the square-shaped terminal. Turn that screw to hold the red wire in place.

3. Twist the black wire around the screw above the round-shaped terminal. Twist that screw to hold the black wire in place.

4. Screw the mini bulb into the bulb receptacle. Flip the switch on the battery holder. The bulb should light up. Congratulations; you made your first circuit!

How it works: The battery holder delivers a charged current to a wire in the bulb. The wire heats up and gives off light. Once the charged current lights the bulb, it loses energy and needs to go back into the battery to get more energy. A battery is like a charge escalator, the moving stairs at a mall. It boosts the low-energy charge back up to a high energy.

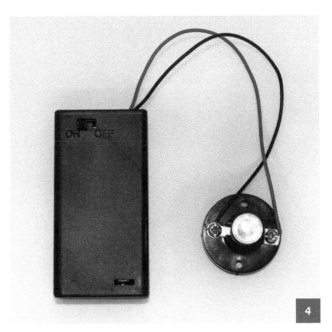

STEAM CONNECTION: An electric circuit is a simple form of technology. Learning the basics of circuits is your introduction into the science of electrical engineering!

BUILD AN LED THROWIE

Now that you've built a simple circuit, let's make an even simpler one we can use for art. An LED, or **light-emitting diode**, was one of the first **semiconductors** invented. A diode is like a one-way street for electric current. An LED gives off **photons**, or particles of light, as a current flows through it. If you connect an LED the "wrong" way around, it will block the current from flowing. Try it!

In the first project, we used a regular bulb that emits light by heating up a wire. LEDs emit light when **electrons**, a type of particle found in atoms (the stuff everything is made of), recombine in the semiconductor in a certain way. This uses less energy than a traditional light bulb, since no heat is needed. Many modern light bulbs, television screens, and jumbo screens use LEDs. Now, we'll build a glowie, or magnetic glow dot.

TOTAL TIME: 5 MINUTES

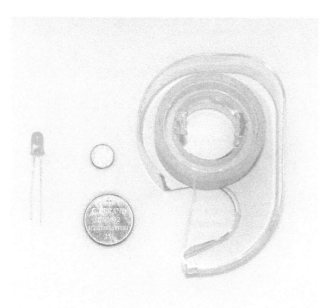

MATERIALS:

- 1 CR2032 3-volt button cell battery
- 1 LED
- 1 small neodymium rare earth magnet
- Cellophane tape

! CAUTION: Don't put magnets or batteries in your mouth, even for a second. You might accidentally swallow them. Batteries are poisonous, and magnets are dangerous if swallowed, potentially magnetizing your intestinal wall and essentially scrambling your insides. Keep magnets away from all televisions, computers, phones, and tablet screens, as magnets can damage screens.

STEPS:

1. Place the button cell battery in between the LED legs. Make sure the smooth, positive (+) side of the battery is touching the longer LED leg. The LED should light up.

2. Use tape to hold the LED snug against the battery.

3. Attach the magnet to either side of the battery.

4. Find different surfaces to which the magnet will stick. Make sure it is somewhere you can reach. You can make many of these LED throwies and use them to decorate your room, as long as you have **ferromagnetic** surfaces for them to stick to.

> **How it works:** Each leg of an LED has a different material at the end that's inside the plastic bulb. Current can flow from one material to the other and give off photons. But it cannot go the other way. LEDs are **polarized**, meaning they only work one way.

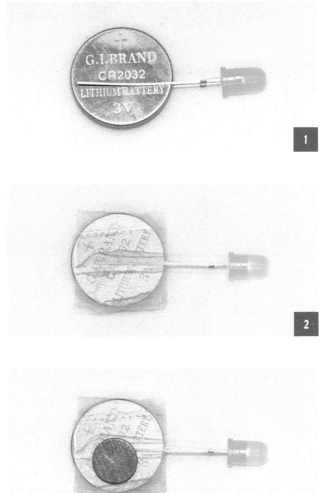

STEAM CONNECTION: These are called LED *throwies* because you can throw them and the strong magnet will stick to steel. Sometimes artists use them to decorate old, rusty metal objects. They can turn ugly structures, such as rusty cars and refrigerators, into colorful and interesting sights. While going around and testing different surfaces with the magnet, you are performing science. LEDs were one of the first types of semiconductor, which is an advanced technology.

BUILD A SERIES CIRCUIT

There are two main types of circuits: series and parallel. First, we will explore **series circuits**. In a series circuit, you connect two components together from their positive (+) to their negative (–) terminals, and so on, back to the battery. Even the batteries are connected in series. When batteries are connected this way, their voltage, or the amount of energy each charge has, adds together. Let us explore series circuits.

TOTAL TIME: 10 MINUTES

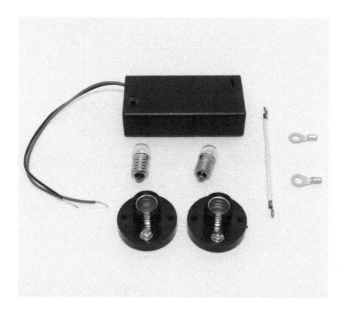

MATERIALS:

- 1 (2-AA) battery holder with on/off switch and wire leads attached
- 2 ring connectors
- 2 mini plastic bulb receptacles
- 2 mini 1.5-volt bulbs
- 1 jumper wire
- 2 AA batteries
- Screwdriver

! CAUTION: Be careful when unscrewing the bulbs. They can get hot if left on for a long time. Ask an adult to help you crimp the wire, as it is *very* easy to accidentally cut yourself on the sharp ends of the wires.

1. Look at the screw terminals on the mini plastic bulb receptacles. On the ones we use for this project, one is square, or positive (+), and the other is round, or negative (–). If your bulb receptacles are different, the negative terminal is connected to the sides where the bulb screws in. The positive terminal is connected to the contact where the bottom of the bulb touches the center. Connect a jumper wire between the positive terminal of one bulb receptacle and the negative terminal of the other. If you do not have a jumper wire with ring connectors on the ends, you can use a regular piece of wire to wrap around the screw.

2. Have an adult help you crimp the ring connectors to the ends of the battery wires. Attach the black wire from the battery to the unused negative (round) receptacle terminal. Connect the red wire to the positive (square) receptacle terminal. Insert the AA batteries. Remember: The flat side of a battery goes against the big spring.

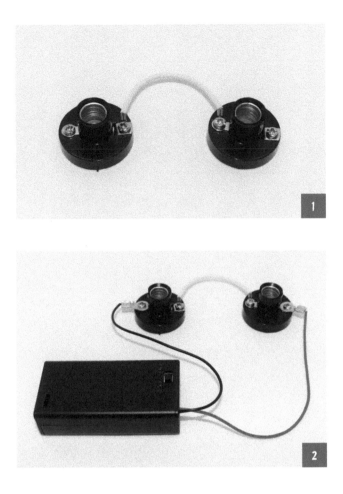

CONTINUED ➡

3. Screw in both bulbs. They should light up. Are they very bright? What happens if you unscrew one of the bulbs?

How it works: In this series circuit, the current passes through one bulb, then the other, then back to the battery. When you unscrew one bulb, you are breaking the circuit, which is why both bulbs turn off. The bulbs glow dimly because they are splitting the voltage of the battery holder.

STEAM CONNECTION: Exploring why and how things work, like you did with the series circuit, is the foundation of science. Light bulbs were one of the first popular forms of electronic technology, and today they are an essential part of everyday life.

BUILD A PARALLEL CIRCUIT

Parallel lines are lines that never touch. Train tracks are an example of parallel lines. Imagine connecting one rail to the positive terminal of a battery, and the other rail to the negative. Now connect one light bulb to each rail, and see it light up. Next connect another bulb to each rail. What will happen? Let's find out by building a **parallel circuit**.

TOTAL TIME: 10 MINUTES

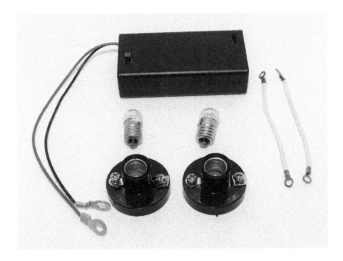

MATERIALS:

- 1 (2-AA) battery holder with on/off switch and wire leads attached
- 2 ring connectors
- 2 mini plastic bulb receptacles
- 2 mini 1.5-volt bulbs
- 2 jumper wires
- 2 AA batteries
- Screwdriver

! **CAUTION:** Be careful when unscrewing the bulbs. They can get hot if left on for a long time.

CONTINUED ➡

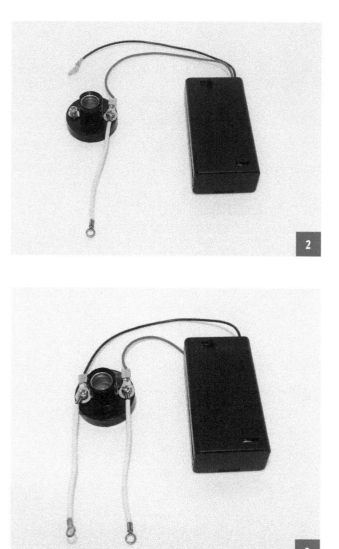

STEPS:

1. Have an adult help you crimp the ring connectors to the ends of the battery wires (or reuse the battery holder from the previous project with the ring connectors already attached—see image 2 on page 25).

2. Look at the screw terminals on the mini plastic bulb receptacles. One is square, or positive (+), and the other is round, or negative (–). Connect a jumper wire *and* the red wire of the battery to the positive (square) terminal of one bulb receptacle.

3. Attach a jumper wire *and* the black wire from the battery to the negative (round) terminal of the same bulb receptacle.

4. Connect the negative jumper wire to the negative terminal of the second bulb receptacle. Then, connect the positive jumper wire to the positive terminal of the second bulb receptacle.

5. Screw in both bulbs. They should light up. Are they very bright? What happens if you unscrew one of the bulbs?

How it works: In this parallel circuit, the current passes through each bulb on its own, then back to the battery. When you unscrew one bulb, the other bulb is still connected to the battery. You could add many bulbs in series, and they would all light up with the same brightness. The bulbs glow bright because they are getting the full voltage of the battery holder.

STEAM CONNECTION: Exploring why and how things work, like you did with the parallel circuit, is the foundation of science. Think about how different our world would be without simple but important inventions like the light bulb.

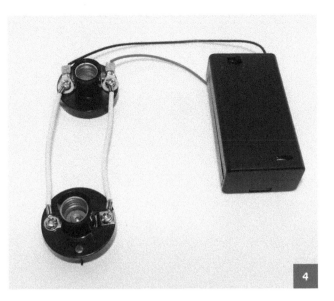

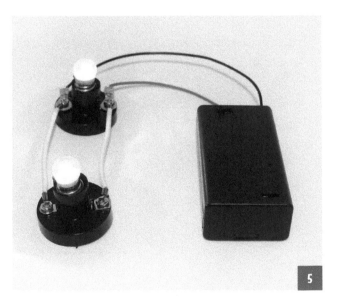

BUILD A ROBOT-LIKE NIGHT LIGHT

Now we will use one of the circuits from the previous projects to power a night light. This night light looks like a robot, but yours doesn't have to look exactly like the one in this book. You can design, or plan out, your own parts for your night light. This project uses the parallel circuit so it glows bright and your room doesn't get too dark, but you can choose to use either kind of circuit.

TOTAL TIME: 45 MINUTES

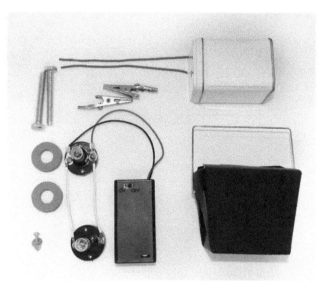

MATERIALS:

- 1 parallel circuit (from the previous projects, with battery holder, jumper wires, bulb receptacles, and bulbs)
- 1 small Chinese takeout container (for the "head")
- 1 small square tin with lid (for the "body")
- 1 small machine screw with washer and nut
- 2 adhesive-backed foam washers
- 2 (¼-inch by 3-inch) bolts (for the "legs")
- 2 (6-inch) pieces of solid core copper wire
- 2 alligator clips
- Hobby knife
- Screwdriver
- Double-sided tape or hot glue gun and glue stick

! **CAUTION:** Be careful when punching holes in the tin, as there could be sharp slivers of metal that could cut you. Have an adult help you cut or punch any holes. Using a hot glue gun also requires caution and supervision as it's possible to burn yourself.

STEPS:

PART A: MAKING THE HEAD

1. Remove any bulbs and place the receptacles upside down on the "face" of your small Chinese food container. Trace the small circle that needs to be cut out so the receptacle shafts can stick through. Cut out the holes you traced with a hobby knife if you feel comfortable using a hobby knife; otherwise, ask an adult to help. Also, cut a very small hole in the bottom for the small screw or have an adult do this for you.

2. Punch out a small hole in the lid, or ask an adult to do this for you. Feed the small machine screw through the hole in the base of the head, from inside, so that the pointy part sticks out the bottom of your container. Slide the lid hole over the screw, and tighten it to the head with a nut.

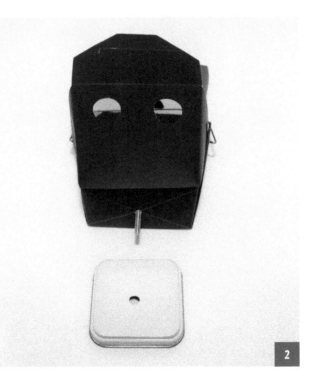

CONTINUED ➡

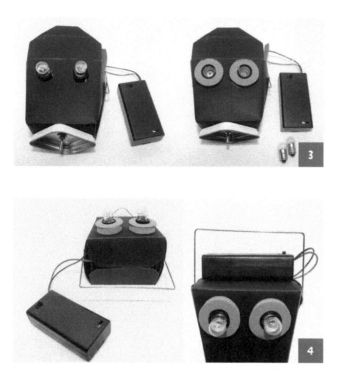

3. Insert the bulb receptacles of the completed and working circuit into the head, and poke the bulbs and receptacle shafts out through the eye holes. Be gentle with the wires from the battery holder. Remove the bulbs, and use the adhesive-backed foam washers to hold the receptacle shafts in place. If you cannot find adhesive-backed foam washers, use some glue, or get creative. Screw in both bulbs.

4. Carefully close the top of the takeout container, or "head," with the battery wires sticking out. Use double-sided tape or hot glue to attach the bottom of the battery holder to the top of the head. Make sure the switch is sticking up.

PART B: MAKING THE BODY

5. Punch a small hole on the top left-hand side of one side of the tin for the wire arm. Punch a larger hole in the bottom center for the leg bolt. If your tin is pretty thick, this may be difficult, so don't be afraid to ask an adult for help. Turn the tin to the right (so that the next side is now facing you), and punch a small hole in the top right for the wire arm and a larger hole in the bottom center for the leg bolt, or ask an adult to do this for you.

6. Make the arms by crimping each alligator clip to each wire. Insert the wires into the small holes on the tin, and twist them together. You can also add a dab of glue on the inside to hold the arms in place, but be careful not to cut yourself, as there may be sharp edges on the inside edge of the holes.

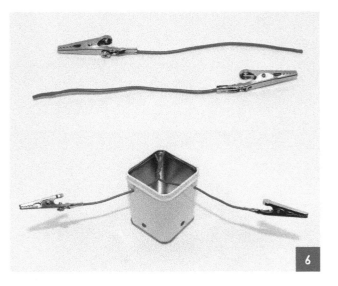

CONTINUED ➡

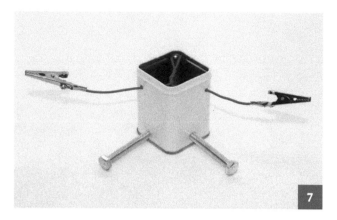

7. Insert the leg bolts, and add a dab of glue to hold them in place.

8. Attach the head and turn on the switch. Your robot night light will protect you from the dark.

9. Get creative to personalize it. Does it need a shirt? Draw one on or make one. What about a smile?

How it works: When you flip the switch, the eyes on the "robot" light up. This project builds on the last one, taking the simple light setup you previously built and housing it in a "robot" body. The alligator clip hands add functionality, allowing your night light to also double as a holder for things like pictures, notes, or other things you might want a good view of in the dark.

STEAM CONNECTION: Designing anything, including your robot night light, takes a combination of art and engineering. The engineering makes the pieces fit and work together. The art makes it look sweet.

Chapter 4

ROBOTS FOR ENTERTAINMENT

Ancient civilizations dreamed of having robots, but they didn't have technology like electricity or computers. They *could*, however, make machines with gears, levers, and **pulleys** that could give the appearance of operating all on their own. People back then would make machines, called automatons, that looked like humans or animals, for entertainment. These automatons could do different things. Some could write words

or draw pictures. Some could sing or play music. Though these robots looked like they were moving on their own, they could only repeat the same movements.

Although automatons were often built for entertainment, they were a good way for inventors to experiment with **mechanics**, hydraulics, and pneumatics, different areas of technology that deal with movement. It was through such playful experimentation that some very useful automatons came to be, such as the cuckoo clock and Hero's engine, a bladeless steam turbine that spins when water in a central container is heated. Many other great inventions throughout history came about because people were willing to take the knowledge available to them at the time and experiment in new and playful ways.

In this chapter, we will honor the original automatons by building some "robots" meant to entertain us, and hopefully, we also find ourselves entertained by the process of building them. We will build a robot that makes art, a robot that dances, and a frog-like robot that jumps around and is hard to catch.

While we build these projects, think carefully about a question we considered earlier in this book—the difference between a true robot and a simple machine. Although you may conclude that some of the "robots" we build aren't robots capable of providing thought-out responses to environmental inputs, remember this: Science builds upon itself. First, people began making automatons. Now think about the early computers that were bulky and had very limited capabilities compared with the computers built today. All of these things were essential building blocks that allowed us to create the high-tech robots in our world now, just like the basic light circuit you built earlier became the foundation for your robot-like night light.

While the "robots" we build in this and other chapters are not very complicated, there are many examples in our world of complex and intelligent robots meant to enhance life by entertaining, including interacting with us in relatable and emotional ways, providing companionship, and chatting with us. By building simple entertainment "robots" now, you are developing the skills to build even cooler ones later.

BUILD A HOMOPOLAR MOTOR

Before we start building robots that move, we need to know *how* robots move. Most robots use electric motors to move. In this project, we will build our own motor. The first electric motor was built by Michael Faraday in 1821. He used a battery, a magnet, wire, and a pool of mercury. We will not be using mercury because it is poisonous. (It's also very expensive!)

We can do the same thing with just a battery, a magnet, and wire. The motors we will use in the other projects are a bit more complex, but they use the same principles. When electric current flows through a wire, it creates a magnetic field. If you bring a **permanent magnet** near the wire, it will oppose that magnetic field and push it away, making the wire move. A motor that runs on direct current and produces a constant circular motion, as we're going to build, is called a **homopolar motor**.

TOTAL TIME: 15 MINUTES

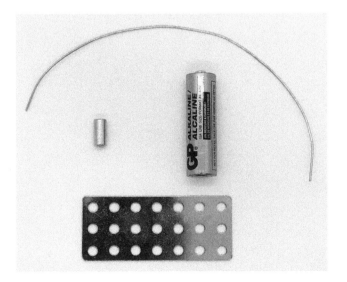

MATERIALS:

- ⊜ 1 AA battery
- ⊜ 1 small neodymium rare earth magnet
- ⊜ 1 piece of solid core copper wire (at least 12 inches, with the insulation removed)
- ⊜ 1 small metal base
- ⊜ Hammer
- ⊜ Phillips head screwdriver, ball bearing, or other small, round object

⚡ CAUTION: If the motor runs for too long, the wire and battery will heat up. Also, have an adult help with the part of the project that involves the use of the hammer.

CONTINUED ➡

STEPS:

1. Place a small ball bearing, or Phillips head screwdriver, on the positive (+) side of the battery (the side with the bump). Very, *very* gently tap with a hammer until a small dent appears. This dent will help hold the wire in place.

2. Bend the copper wire into an "M" shape. The size should be close to the size of the battery. The bottom of the wire should bend to touch the magnet or the bottom (negative side) of the battery.

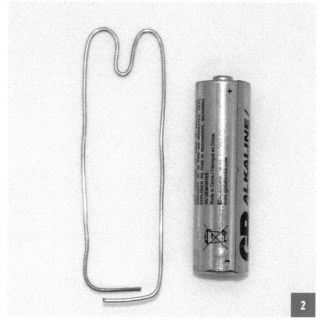

3. Place the magnet, standing upright, on top of the metal base.

4. Stand the battery up on top of the magnet with the negative (–) or flat side of the battery pointing down, touching the magnet, and the positive (+) side facing up.

5. Place the downward point of the wire (the middle of the "M" shape) in the dent at the top of the positive (+) side of the battery. The wire should balance on the battery.

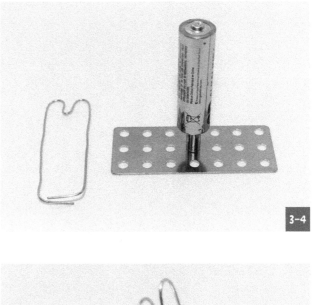

3–4

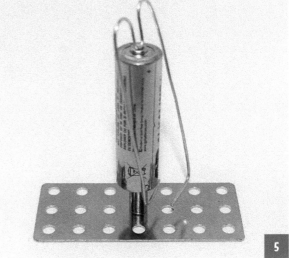

5

CONTINUED ➡

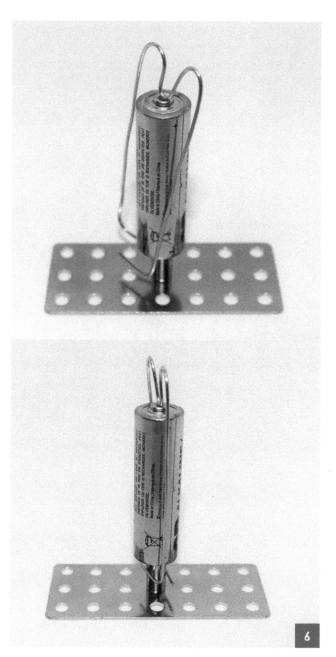

6. Put one "leg" of the wire "M" on each side of the magnet. The wire should start spinning around.

How it works: When an electric current flows through a wire, it creates a magnetic field. If you bring a permanent magnet near the wire, it will oppose that magnetic field and push it away, making the wire spin.

STEAM CONNECTION: Using the properties of electricity and magnets to make the wire spin is science. A motor is a type of technology. Bending the wire into the right shape uses engineering.

BUILD A SCRIBBLER BOT

Have you ever gotten a pizza delivered and noticed that it has a little three-legged plastic table in the middle? That's called a pizza saver, and it keeps the paper or cardboard from sticking to the melted cheese so you can enjoy the pizza without eating the box, too. The scribbler bot uses a pizza saver as a "found object" for the base of the robot. Reusing common household or discarded items, also known as **found objects**, in different ways is a great skill for both art and robot making. Finding creative uses for what some may think of as junk is an important skill for robot builders. As the name implies, the scribbler bot can scribble for you. The question is: What will it scribble? You'll have to build it to find out.

TOTAL TIME: 30 MINUTES

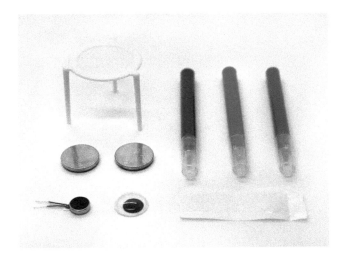

MATERIALS:

- 1 pizza saver tripod
- 2 CR2032 3-volt button cell batteries
- 1 vibrating disk motor
- 2 glue dots
- 3 small markers
- 1 googly eye
- Cellophane tape or vinyl electrical tape

⚠ **CAUTION:** The wires on the vibrating disk motor may have very small amounts of the metal wire exposed. You need between ¼ and ½ inch of wire for best results. Carefully scrape away the plastic insulation with a hobby knife if needed, but be careful, or ask an adult for help. The wires are thin and fragile, and you don't want to accidentally cut yourself.

CONTINUED ➡

STEPS:

1. Remove the sticker paper backing from the vibrating disk motor, and place the motor adhesive side down on top of the pizza saver tripod. It does not need to be perfectly centered. Try placing it off center and see what happens.

2. Place a glue dot on top of the disk motor, and bend the black wire so the metal tip is just barely touching the glue dot.

3. Stack the two CR2032 batteries in series, so both positive (+) sides are facing up, and tightly tape them together. Cellophane or vinyl electrical tape works well for this.

4. Place this battery stack negative (–) or rough side down on the glue dot, making sure the negative (–) side of the battery touches the metal tip of the motor wire. Bend the red wire over the top, and make sure it easily touches the positive (+) side of the battery. You should feel a buzz.

5. Flip the scribbler bot over. Tape the three markers onto the legs of the tripod, making sure the tips of the legs come right up to the caps of the markers.

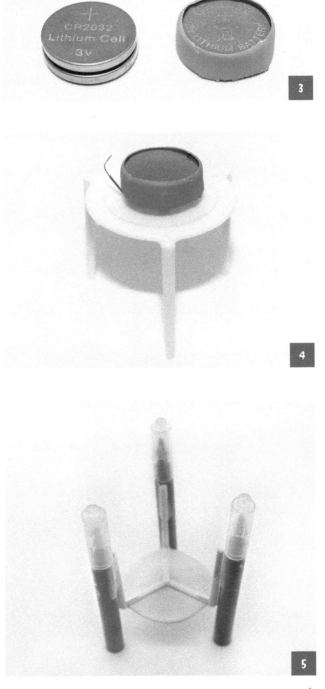

CONTINUED ➡

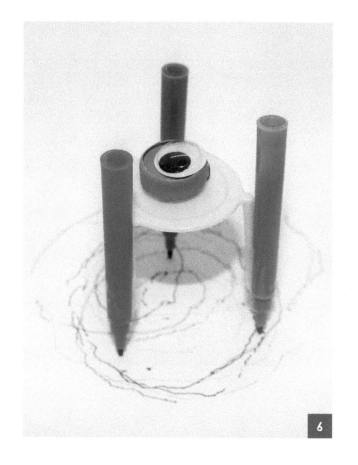

6

6. Flip the scribbler bot back over. Stick the googly eye to a glue dot. Use the sticky side of the glue dot to stick the red wire to the positive (+) side of the battery. You should feel it buzzing. Take the marker caps off, and put the scribbler bot on a piece of paper. It should spin around and make interesting art.

How it works: The motor powers the scribbler bot, making it move around on its own. How it moves depends a lot on the placement of the motor atop the pizza saver tripod. If you remove the caps from the markers and let your scribbler bot go, you'll end up with an interesting pattern of lines. Just be sure to set the scribbler bot down on paper, not a countertop where the markers may cause unwanted "art."

STEAM CONNECTION: Creating a structure with legs that can move around to scribble is engineering. Connecting your motor to the batteries is technology. And the end product is a tool for making art. Experimenting with the placement of the motor on the pizza saver tripod to achieve different kinds of scribbles is science.

BUILD A SOLAR DANCING ROBOT

The scribbler bot from the last project makes visual art, or still art. Now we'll build a robot that makes kinetic art, or moving art. Dancing is fun exercise, but it is also a form of art. Dance can tell a story through movement. By combining different moving objects on this robot, you can make it tell a story through movement, too.

This robot uses a small solar panel as a power source. Solar panels turn light into electricity. The solar panel used for this project has springs that you can connect to the wires from the motor. If you cannot find this same type of solar panel, that's okay. Simply find the smallest one you can. This robot works best in the sun, but it can also work with a very bright light. Try it outside, then test it with lights in your house.

TOTAL TIME: 30 MINUTES

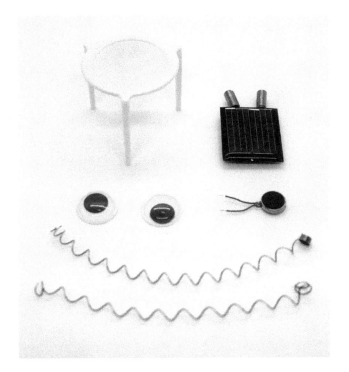

MATERIALS:

- 1 small solar panel, rated 0.5V, 800mA, or higher
- 1 vibrating disk motor
- 1 pizza saver tripod
- 2 googly eyes
- 2 springs
- Double-sided cellophane tape or a glue dot
- Other decorations that jiggle (optional)

! CAUTION: The wires on the vibrating disk motors may have very small amounts of the metal wire exposed. You need between ¼ and ½ inch of wire for best results. Carefully scrape away the plastic insulation with a hobby knife if needed, but be careful, or ask an adult for help. The wires are thin and fragile, and a knife can cut you.

CONTINUED ➡

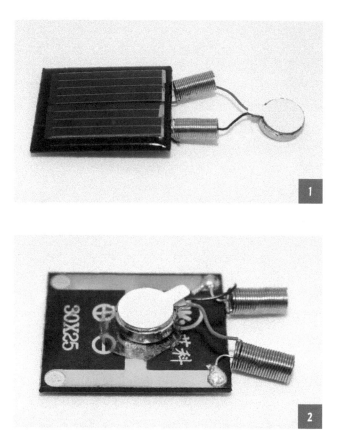

STEPS:

1. Connect the leads, or wires, from the vibrating disk motor to the solar panel. If your solar panel has spring attachments, pull the spring out, slip the wire into the middle, and let the spring close around the wire. If your solar panel has wires, just twist the metal parts of the wires together.

2. Use some double-sided tape (or a glue dot) to connect the top of the motor to the bottom of the solar panel.

3. Remove the paper backing of the vibrating disk motor to expose the sticky side. Stick the motor and solar panel to the pizza saver tripod. Now decorate your dancing robot with googly eyes, springs, and other things that jiggle. Take it into the sun (or a bright light), and watch it dance.

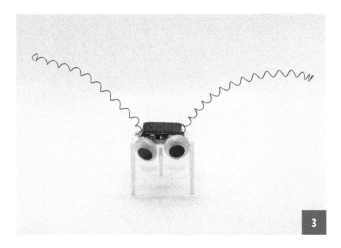

How it works: The solar panel turns light into electricity. The electricity then powers the small vibrating motor, which makes everything jiggle and dance.

STEAM CONNECTION: When you test your robot under different light sources, you are using science to discover what makes your robot work. You use art to decorate your robot. Making different materials wiggle uses engineering.

BUILD A JUMPING FROG ROBOT

I don't know if you have ever tried to catch a frog in a creek, but I think that should be an Olympic sport. Frogs are fast and slippery, and they jump around, which makes them fun to chase. But we don't want to hurt a frog; it's just fun to run around and try to catch one. In this next project, we will build a jumping frog-like robot that you and your friends can chase around and try to catch.

TOTAL TIME: 30 MINUTES

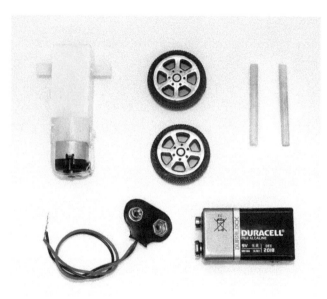

MATERIALS:

- ➲ 1 dual shaft gear motor
- ➲ 2 small plastic wheels with spokes
- ➲ 1 (9-volt) battery snap connector
- ➲ 1 (9-volt) battery
- ➲ 2 (1½-inch) wooden rods, ⅛ to ³⁄₁₆ inch in diameter
- ➲ 2 googly eyes
- ➲ Double-sided foam tape
- ➲ Hot glue gun and glue stick

! **CAUTION:** Be careful when using hot glue, as it gets *really* hot and could burn you. Ask an adult for help. Do not touch it until it is fully cooled.

STEPS:

PART A: ATTACHING THE WHEELS

1. Turn the wheel backward, and fit the shaft through one of the spokes. The wheels used in this project have spokes through which the gear motor shaft fits snugly. If you can't find the same wheels, that's okay. The important part is that the wheels are off center. Make your wheels work.

2. Fit the second wheel to the other shaft. Make sure the wheels are equally offset and that they line up.

3. Apply some hot glue to connect the motor shaft to the wheel. Repeat on the other side. Make sure you do not glue the shaft to the motor gearbox.

CONTINUED ➤

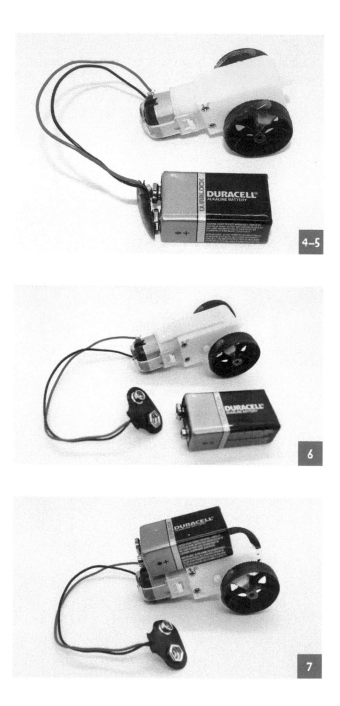

PART B: TESTING THE MOTOR/ FINDING FORWARD

4. Connect the wires from the 9-volt battery snap to the motor terminals. Hold the gearbox motor up while you connect the battery. Notice which way the wheels spin. You want the wheels to propel the robot in the direction of the wires connected to the motor. If it is going the opposite way, flip the motor around.

5. Add traction by squeezing a line of hot glue around *only half* of the outside of each wheel. Start where the gearbox shaft is closest to the edge, and move in the direction opposite from how the wheel spins. This will help the robot jump. Test the robot. If it doesn't jump well, take off the wheel glue and put it on the other half of the wheel. Experiment and see what works best.

PART C: COMPLETING THE BODY

6. Put a strip of double-sided foam tape on the top of the gearbox, for sticking on the battery.

7. Attach the battery to the top of the gearbox. Place the battery as far forward and close to the wires as possible.

8. Attach the front legs by gluing one wooden rod on each side of the motor, **perpendicular**, or at a right angle, to the gearbox.

9. Connect one terminal of the battery snap to the 9-volt battery. When you're ready, connect the other one, and the robot frog will take off and jump away. Don't forget to add googly eyes.

How it works: Adding glue to half of each wheel allows the robot to jump every time that half comes in contact with the ground, and to move forward when the other half does. If you added glue around the whole wheel, the robot would immediately flip over and roll backward.

STEAM CONNECTION: Testing which way to glue your wheels uses science and engineering. Figuring out how far half a wheel is uses mathematics.

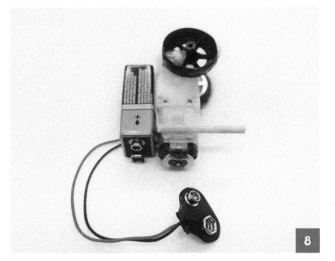

Chapter 5

ROBOTS IN SPACE

Robots are very important to understanding what happens in space. In fact, much of what we know about space actually came from robots sending information, or data, back to Earth. Many space probes and human-made satellites are types of robots. But why do we send more robots to space than humans?

1. **It's safer.** Space is a harsh environment. It gets really cold and really hot. There is **radiation**, which is poisonous to humans. Robots can be made to withstand the harsh environment of space for a long time. Humans are fragile.

2. **It's easier.** Humans need many different things to stay alive. We need oxygen to breathe, water to drink, food to eat, and a place to relieve ourselves. And everything humans need has to be sent up to space with them. Robots, on the other hand, just need electricity. It is easier to generate electricity for a robot than to create a place for people to eat and go to the bathroom in space.

3. **Robots last longer.** Humans can last in space for a while, but at some point they have to come back to Earth. Robots can stay out in space for a long time.

The longest a human has ever spent in space was 437 days in a row, a record set by Valeri Polyakov of Russia. The Voyager 2 probe, launched on August 20, 1977, has spent more than 40 years in space and is still sending data to NASA.

There are many different types of robots in space, but all are meant to help humans in some way, whether by collecting samples in faraway, rough terrains of other planets, taking measurements that would be very difficult or impossible for humans to do, or fixing and assembling equipment. Remote-operated vehicles like NASA's Curiosity can send back photographs and data about planets like Mars, which humans can't yet explore in person. This data gives us important clues about planets' potential habitability. NASA's Remote Manipulator System, part of the International Space Station (ISS), is a robot arm that can

perform remote assembly as well as position and anchor structures. Robots like this are important because they free up astronauts to work on important research and carry out experiments, and they keep humans from doing tasks that could put them at risk. Robonaut is a NASA-built humanoid robot that lives on the ISS, carrying out some of the tasks necessary to keep the station orbiting. The ISS could even be considered a type of space robot, as it performs many tasks on its own to support life.

In this chapter, we will make robots that are modeled after spacecraft, even using some of the same parts as space robots, like solar panels and light sensors. Space robots are often useful not just in space but right here on Earth, for exploring difficult-to-reach areas, taking pictures, and collecting soil samples. Can you think of any ways that the robots we build could provide help to astronauts in space?

BUILD A SOLAR-POWERED ORBITER MOBILE

The International Space Station and many space probes, Mars landers, and satellites are powered by solar panels, which turn sunlight into electricity. In this project, we will use solar panels to directly run a motor. For this project, it is important to find a low-voltage, low-current motor.

We're going to make a solar-powered spaceship **mobile** that will balance on a bottle and spin around, orbiting the bottle. This spaceship is modeled after the Russian spacecraft Soyuz. Riding on the Soyuz is currently the only way (as of this writing) for astronauts to get to the International Space Station. You can make your model based on any spaceship you want.

TOTAL TIME: 45 MINUTES

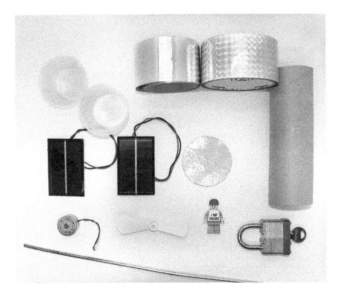

MATERIALS:

- 2 small solar panels, rated 0.5V, 800mA, or higher
- 1 low-voltage, low-current direct current (DC) motor (with wires)
- 1 propeller
- 1 cardboard tube
- 2 or 3 plastic sauce containers (from local restaurant)
- 1 small circle of thick paper (like from a paper plate) covered in aluminum foil
- 1 counterweight (this project uses a small padlock)
- 1 or 2 rolls of aluminum foil tape (for decoration)
- 1 (2½-foot) piece of wire clothes hanger
- 1 plastic bottle (with dimple in its cap)
- 1 small robot or astronaut figurine (about 1½ inches tall, such as a LEGO minifigure)

CONTINUED ➡

- Hot glue gun and glue stick
- Hobby knife
- Marker
- Scissors
- Vinyl electrical tape
- Other materials for decorating the spaceship (optional)

CAUTION: Be careful when using the hot glue gun and hobby knife. Ask an adult for help. Finding a propeller that fits on the motor shaft may be tricky. You might have to hot-glue the fan on or use tape. Make sure when the propeller spins, it is spinning the motor shaft.

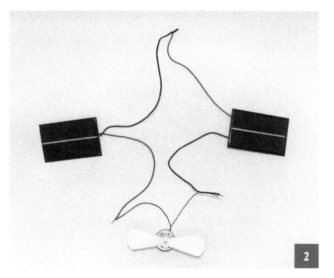

STEPS:

PART A: TESTING THE SOLAR CIRCUIT

1. Make sure the solar panels make the motor and propeller spin. Connect the solar panels in series to power the motor. Twist the red wire of one solar panel to the black wire of the other solar panel.

2. Twist the remaining two wires to the motor, matching red to black and black to red. Attach the propeller, or fan, to the motor shaft.

3. Take the circuit out in the sun to test it. The fan should spin. It is important to make sure the fan is blowing air at you. If not, switch the motor wires to the opposite solar panel wires, and the fan will spin in the opposite direction.

PART B: CREATING THE SPACESHIP

4. Glue the robot or astronaut figurine to the inside of one sauce cup, facing the bottom of the cup. Then, glue the edges of the two sauce cups together to form a capsule.

5. Cover the cardboard tube in aluminum foil tape so it looks like a spaceship. Leave 1 inch or so of foil tape over the back end of the paper tube so you can tape the motor and propeller to it later on.

6. Glue or tape the capsule onto the front end of the tube.

CONTINUED ➡

PART C: ATTACHING THE SOLAR PANEL "WINGS"

7. Find the horizontal centerline of the tube (keep the figurine sitting or standing upward to help). Near the back end of the tube, mark the length of the short side of the solar panel on the centerline. Do this on both sides of the tube. Use a hobby knife to cut a slit along the lines, through the tube, or ask an adult to do this for you. You'll need slits on both sides.

8. Disconnect the motor wires from the solar panels. Carefully feed the solar panel wires through the slits and out the back of the tube. (Be very careful so the wires don't break off.)

9. Insert the solar panels into the slits on each side of the tube. Make sure the solar panels both face up in the same direction. You can glue the solar panels along the slits if they do not fit snugly enough. Make about eight cuts with scissors into the leftover foil tape on the back end of the tube, and fold them back to look like a flower.

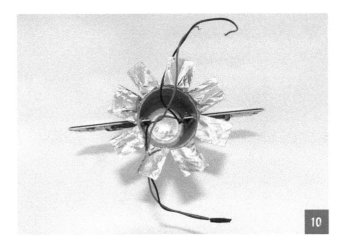

PART D: RECONNECTING THE CIRCUIT

10. Twist together the red wire from one solar panel to the black wire from the other solar panel. Cover the twisted wires with vinyl electrical tape.

11. Stuff the connected wires into the tube.

12. Connect the motor wires to the remaining solar panel wires. Make sure everything is connected the same way as when it was tested.

PART E: MOUNTING THE MOTOR

13. Cut a slit in the paper circle from the edge to the center, or along the radius of the circle. Put the motor on top of the circle, and feed the motor wires through the cut.

CONTINUED ➡

14. Fold one edge of the cut over the other, forming a cone. Tape or glue the overlapping fold to keep the cone shape.

15. Center the motor inside the cone. Make sure the propeller doesn't hit the cone. Glue the motor in the center of the cone.

16. Stuff the motor wires into the back end of the tube, and place the cone and motor over the opening. Fold the extra strips of foil tape over the cone to hold it in place.

17. Place some tape on the underside of the solar panels to hold the wires in place. If you have shiny gold tape, use it, as it looks extra spacey and cool. This is a good time to decorate your spaceship with stickers, antennas, satellite dishes, etc.

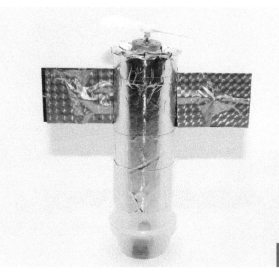

PART F: ASSEMBLING THE MOBILE

18. Straighten the 2½-foot length of clothes hanger wire. Make a loop at one end. Bend about 5 inches of wire into an "L" at the other end.

19. Use a hobby knife to cut a hole in the top and bottom of your spaceship, just in front of the solar panels, or ask an adult to do this if you're not comfortable using a hobby knife.

20. From the top side, feed the bent "L" section of the wire through both holes. Once through, bend about a 1-inch-long section along the spaceship's bottom centerline, and tape it to the bottom of your spaceship.

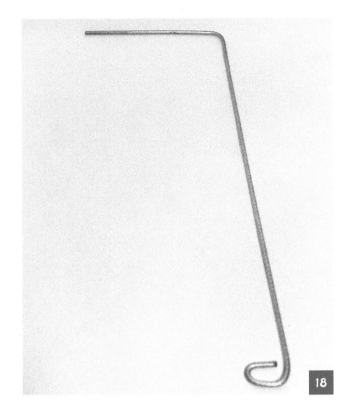

CONTINUED ➡

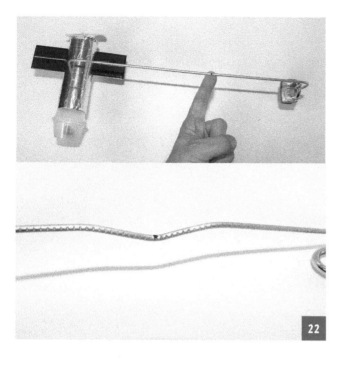

21. Put the small padlock through the loop on the other end. (It helps if the lock or counterweight is slightly heavier than the spaceship.)

22. Use a finger to find the **center of gravity** so both ends of the wire balance evenly. Mark the center of gravity with a marker, then bend the wire at that point into a small "V" with the marked point at its center.

23. Find a plastic bottle with a dimple in its cap. If you cannot, then create a small dimple with scissors or a hobby knife. This will help keep the mobile centered on the bottle cap. Fill the bottle with water so it doesn't fall over, and balance the mobile on the bottle by placing the point of the "V" in the cap's dimple. Set your mobile up in the sun, and watch it spin around in orbit.

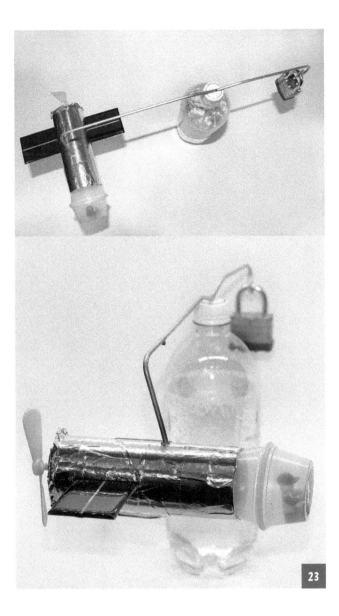

How it works: The solar panels turn photons, or light particles, into electricity. This electricity powers the motor. The motor spins the propeller, pushing air backward and pushing the spaceship forward.

STEAM CONNECTION: Albert Einstein, a famous scientist, first explained how light can be converted into electricity. Solar power is an important form of clean energy technology. Finding the center of gravity on the mobile uses engineering and mathematics. Decorating your solar mobile like a spaceship is art.

BUILD A PB-D2 DROID

One of the most popular robots from movies is R2-D2. R2-D2 is a "droid," which is short for **android**. Androids are robots that act like humans. R2-D2 does not look like a human, but it can talk to humans, hack into space stations, and annoy its friend, C-3PO, which are all things that humans can do.

This robot is called PB-D2, because it looks like R2-D2, but its body uses an empty peanut butter jar. Make sure you clean the jar very well. You don't want any peanut butter gunking up your circuits. After you build PB-D2, you can decorate it like a droid. You can even use LED throwies like the one on page 22 to light it up.

TOTAL TIME: 45 MINUTES

! **CAUTION:** Have an adult help you cut the holes in the plastic jar to avoid accidents. The bottom of a jar is usually thicker than its sides. This project also requires using hot glue, which calls for extra caution so as not to get burned; have an adult help you with the glue as well.

MATERIALS:

- 1 (16-ounce) plastic jar, emptied and cleaned
- 1 (2-AA) battery holder with on/off switch and wire leads attached
- 2 AA batteries
- 1 dual shaft gear motor
- 2 wheels (that fit on the gear motor oval shafts)
- 2 thin plastic wheels (for the back, balancing legs)
- 2 small bolts with washers and nuts (lock nuts, if possible; for the back-wheel **axles**)
- 2 (6-inch) craft sticks
- 1 small plastic sauce container (optional)
- 1 small domed lid
- 2 small screws
- Hot glue gun and glue stick
- Hobby knife
- Marker
- Paint (optional)

STEPS:

PART A: MAKING THE BODY

1. Place the battery holder on the side of the jar, and trace around it with a marker. Use a hobby knife to cut out this rectangle, or ask an adult to do it for you. You want the battery holder to fit in this opening very snugly. That way, it will be easy to pull it out to change the batteries. If the opening is too big, that's okay; just add some hot glue or tape so the battery holder stays put.

2. Place the jar upside down on a table. Hold the gear motor upright, with the motor side facing down on the center of the jar bottom. Trace around the motor. The bottom of a jar will likely be thicker than its sides. Get an adult to carefully cut out this hole with a strong knife, as a hobby knife may not do the trick in this case.

3. Attach the wheels to the gear motor shafts using two small screws.

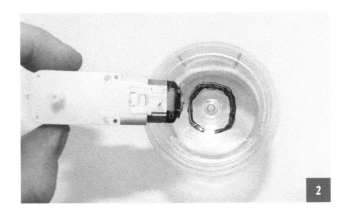

CONTINUED ➡

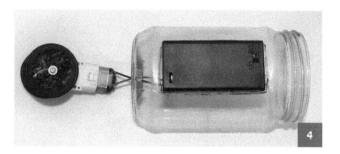

4. Place the battery holder in the side opening so the switch is sticking out. Feed the battery holder wires through the hole at the bottom of the jar. Carefully twist the ends of the wires to the metal tabs on the motor. Be very gentle, because the motor tabs are usually fragile.

5. Test the circuit by turning on the switch. Note which way the wheels spin. You want them to move forward toward the battery holder side of the jar. (The battery holder is the front "belly" of the robot.)

6. Carefully insert the gear motor into the hole cut in the bottom of the jar. Make sure you do not snag the wires on the hole or tear off the motor tabs.

7. Test the motor by turning on the battery holder switch. You want the battery holder side to move forward. If it moves backward, carefully spin the motor around.

PART B: MAKING THE LEGS

8. Cut 1 inch off the 6-inch craft sticks using your hobby knife, or ask an adult to do this for you. Next, make a hole ½ inch in from the new, square end of the sticks. The hole should be wide enough to fit the small bolt you'll be using to attach the back, balancing wheels.

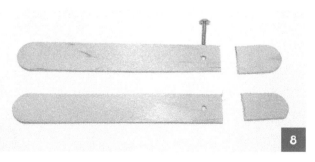

9. Insert one bolt into one of the thin wheels, add a small washer on the other end, then insert the bolt into the hole on the leg. Secure it with a lock nut if you have one, or two regular nuts tightened together. This will keep the nuts from unscrewing as PB-D2 rolls around. Repeat for the other leg.

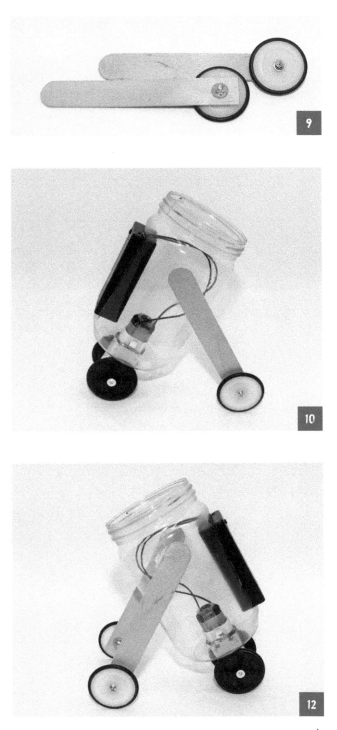

10. Attach the left leg by holding one leg against the left side of the robot with the wheel facing outward; angle the leg back about 60 degrees, or until it looks right. Mark the position of the leg. Squirt a glob of hot glue onto the rounded end of the leg, on the opposite side of the wheel, being careful not to get burned. Ask an adult for help if you're not used to working with hot glue. Press the glued end to the mark. Hold until the glue sets.

11. Once the glue has set, stand the robot up. It should stay up on the three wheels. Make sure it doesn't tip forward easily. If it does, angle the leg further back and re-glue.

12. Hold the right leg up to the other side, and make sure it mirrors, or is at the same location and angle as, the left leg. Make sure all four wheels touch the table or work surface. Mark the position of the right leg with a pencil or marker. Squirt a glob of hot glue on the rounded end of the leg, on the opposite side of the wheel. Press the glued end to the mark you made. Hold the leg in place until the glue has set.

CONTINUED ➡

13

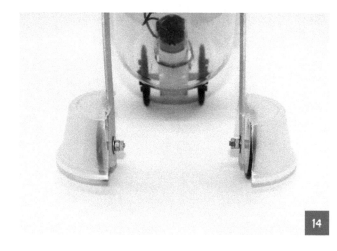

14

PART C: PUTTING ON YOUR FINISHING TOUCHES

13. Glue the dome to the lid of the jar, then screw the lid onto the jar. Your robot should look like a cool droid now! Turn on the switch, and watch it roll forward.

14. (Optional) Cut a small plastic sauce cup in half, and glue each half to the outside of the back wheels. This protects the wheels and also makes it look more like R2-D2.

15. (Optional) Decorate your droid. Paint it white and blue to match R2-D2, brown to more closely resemble its peanut buttery beginnings, or with your own artistic design!

How it works: The motor moves PB-D2 forward, while the back legs keep its balance.

STEAM CONNECTION: You used mathematics and engineering when you were finding the right angle of the legs. Decorating PB-D2 required your artistic skills. This robot is based on a fictional robot from a science fiction movie. Science fiction combines art and science to try to tell a futuristic story.

BUILD A BREADBOARD LED CIRCUIT

A breadboard is a really cool base for making different circuits. When people first started experimenting with electronics, they needed something to practice with and test circuits. People put nails in a wooden breadboard (used for cutting bread). Then, they would twist the electrical components between the nails to make their circuits. Now, you can buy breadboards made of plastic that the electrical components plug into.

In the next two projects, we will explore how breadboards work. We will make some simple circuits, or **prototypes**, to practice our circuit-making skills. Breadboards have many rows of holes to plug in electronic parts. A gap separates each row. The rows on each side of the gap all connect together. In the breadboard used in this project, for each row, holes ABCDE connect together, and holes FGHIJ connect together. In this project, we will light up an LED.

TOTAL TIME: 15 MINUTES

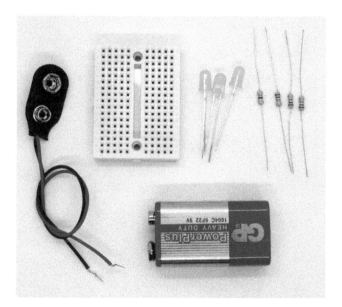

MATERIALS:

- 1 small, 170-point breadboard
- 3 LEDs
- 1 (9-volt) battery
- 1 (9-volt) battery snap connector
- 1 (470-ohm) resistor
- 3 (150-ohm) resistors
- 3 jumper wires (any short length of solid core wire)

! CAUTION: Do not connect an LED directly to a 9-volt battery. Always use a resistor, which reduces the flow of electrical current, so you don't get shocked by accident.

STEPS:

1. Plug the long lead of one LED into A1, and the short lead into B1.

2. Plug one side of the 470-ohm resistor (marked with yellow, purple, and red stripes) into C2, and the other side into C6.

3. Connect the 9-volt battery to the 9-volt battery snap.

4. Plug the metal tip of the red wire into E1.

5. Plug the metal tip of the black wire into E6. The LED should light up. Make sure the long leg of the LED is in A1 and the short leg is in B1. If the LED is backward, it will not work.

6. Replace the 470-ohm resistor with three 150-ohm resistors (marked with red and green stripes). In the picture, the first connects between B2 and B6, the second connects between C6 and C10, and the third connects between D10 and D14.

CONTINUED ➡

7. Now try replacing the resistors with LEDs, using jumper wires (if needed) to complete the circuit. Look at how the brightness changes or stays the same. What happens if you use one LED and two 150-ohm resistors?

How it works: A breadboard allows you to make quick connections, so you can swap out electrical components easily to perfect a circuit.

STEAM CONNECTION: A breadboard is an important tool for developing new technologies. Trying different combinations of resistors and LEDs is science. Finding the different resistor values uses mathematics.

BUILD A SPACE SOUND EFFECT ROBOT

This project is based around the classic 555 timer integrated circuit, or IC chip, which is used in many applications. The 555 timer measures ¼ inch by ½ inch but contains 25 **transistors**, 2 diodes, and 15 resistors. This complex circuit is simplified down to eight connections. This allows us to easily build complex projects with a small chip.

In this next project, we will make a space sound device. It uses a light-sensitive resistor as the sensor, and it emits a different sound depending on how much light it detects.

TOTAL TIME: 35 MINUTES

MATERIALS:

- 1 small, 170-point breadboard
- 1 555 timer integrated circuit
- 1 (9-volt) battery
- 1 (9-volt) battery snap connector
- 1 photoresistor
- 1 (1,000-ohm) resistor
- 1 (1 million-ohm) resistor
- 1 (1-μF) capacitor
- 1 (100-μF) capacitor
- 1 momentary push-button switch
- 4 jumper wires (any short length or color of solid core wire)
- 1 small speaker with wires attached

! CAUTION: Make sure you insert the **capacitors** the correct way. Some electrical components, like resistors, can be connected either way. But **electrolytic** capacitors only work when connected one way. In this circuit, if you connect a capacitor backward, the circuit will just not work. But in future electronic projects you may build, connecting capacitors backward might cause them to burst or damage other components.

CONTINUED ▶

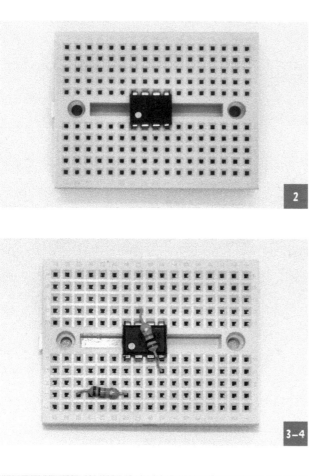

STEPS:

1. Orient the breadboard so that the A1 contact point is in the top right corner.

2. Place the 555 IC chip with the notch facing left, between E8 and E11, and F8 and F11.

3. Place the 1,000-ohm resistor (marked with brown, black, and orange stripes) between G8 and D10.

4. Place the 1 million-ohm resistor (marked with brown, black, and green stripes) between I15 and I11.

5. Place the 1-µF capacitor's long, positive (+) lead in H11, and its short, negative (–) lead in H15.

6. Place the 100-µF capacitor's short, negative (–) lead in J9, and its long, positive (+) lead in J2.

7. Place one jumper wire (orange in image) between H8 and D11.

8. Place one jumper wire (yellow in image) between D9 and F15.

9. Place a third jumper wire (brown in image) between E7 and G11.

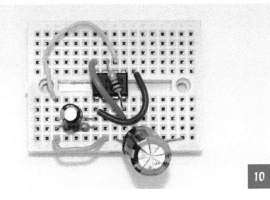

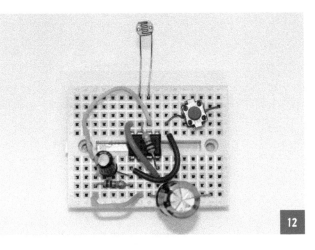

10. Place one jumper wire (green in image) between J10 and J15.

11. Place the push-button switch between B1 and D7.

12. Place the photoresistor between A9 and A10.

13. Remove the adhesive backing from the bottom of the breadboard. Attach the skinny side of the 9-volt battery to the side under rows 13 to 17, with the terminals on the A–E side. Attach the metal end of the speaker to the remaining adhesive area. Flip the breadboard back over.

14. Connect the red wire from the speaker to I2.

15. Connect the black wire from the speaker to E1.

16. Snap the 9-volt battery connector to the battery terminals.

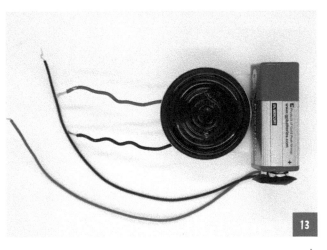

CONTINUED ➡

17. Connect the red wire from the battery connector to I8.

18. Connect the black wire from the battery to A1.

19. Press the switch to activate the oscillator, emitting a tone from the speaker. Cover the photoresistor, and the tone should get lower in pitch. Uncover and shine a flashlight on the photoresistor to make old movie UFO sounds.

How it works: The 555 timer sends steady pulses of electric current. When the IC chip is connected to a speaker, those pulses produce blips. When the resistance value changes, the frequency, or speed and number, of the blips changes, altering the sound emitted from the speaker. To change the frequency, we use a special type of resistor called a photoresistor, which changes its resistance depending on the intensity of light it detects.

STEAM CONNECTION: A 555 timer is an integrated circuit, which is an advanced form of electronic technology. Breadboards are an important tool to develop new circuits and use science. Trying different combinations of resistors and capacitors is a form of experimentation, which is also science. Finding the different resistor values uses mathematics.

BUILD A GRAVITY-DEFYING ROBOT

If you think about it, any robot that moves defies gravity. Even just standing up is technically defying gravity. But this robot defies gravity in an awesome way. It has magnetic tape on its bottom and can walk up magnetic surfaces, like your refrigerator or a metal door. When traveling horizontally, it can even walk over uneven surfaces.

TOTAL TIME: 1 HOUR AND 45 MINUTES

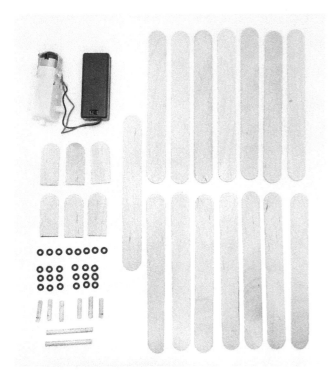

MATERIALS:

- 1 (2-AAA) battery holder with on/off switch
- 2 AAA batteries
- 1 dual shaft gear motor
- 15 (6-inch) craft sticks
- 6 (1½-inch) pieces of craft sticks
- 2 rubber bands
- 2 (1¾-inch) wooden rods, ⅛ to 3⁄16 inch in diameter
- 6 (¾-inch) wooden rods, ⅛ to 3⁄16 inch in diameter
- 26 small rubber O-ring gaskets, ⅛ inch in inner diameter (must fit snugly on wooden rods)
- 2 small screws with washers
- 3 (5-inch) strips of adhesive-backed magnetic tape
- Small screwdriver
- Drill
- Hobby knife
- Hot glue gun and glue stick

⚠️ **CAUTION:** Be careful when using hot glue, as it gets really hot and could burn you. Ask an adult to help you. Do not touch the glue until it is fully cooled. Ask an adult to drill the holes for you.

CONTINUED ➡

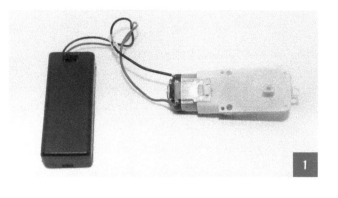

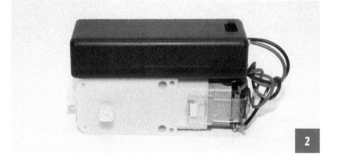

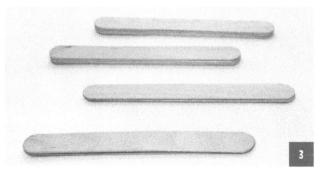

STEPS:

PART A: ASSEMBLING THE BASICS

1. Insert the wires from the battery holder into the holes in the metal tabs on the motor. Carefully twist to secure them, making sure not to break the motor tabs.

2. Glue the battery holder, switch side up, on top of the motor. Tuck any loose wires between the end of the motor and battery holder.

3. Hot-glue two sets of four craft sticks together. Then, hot-glue two sets of two craft sticks together.

4. Once the glue has completely cooled, stack the craft sticks in a pile, and wrap them together tightly with rubber bands. Make a mark directly in the center of the top craft stick. Make additional marks ½ inch from each end.

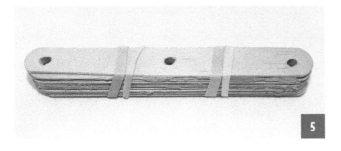

5. Using a drill bit just wider than the wooden rods, ask an adult to start at each mark and drill straight down through all the craft sticks, making sure they stay tightly together. You want each stick to have holes in the same place. If your adult has a drill press, that is a great tool to use for this step.

6. Using the same drill bit, have the adult drill a hole ½ inch from each end of the small craft stick pieces.

7. Attach an O-ring about ⅛ inch from one end of each of the six short wooden rods.

8. Apply glue around the end of the shafts of the six short wooden rods by the O-rings, and press those ends into the holes at the rounded ends of the short craft stick pieces. Make sure the ends of the shafts are flush with the bottom of the short craft sticks and do not stick out.

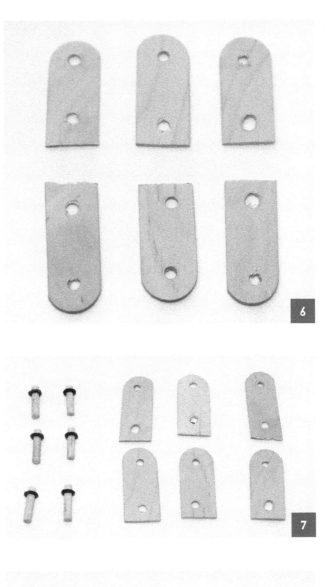

CONTINUED ➡

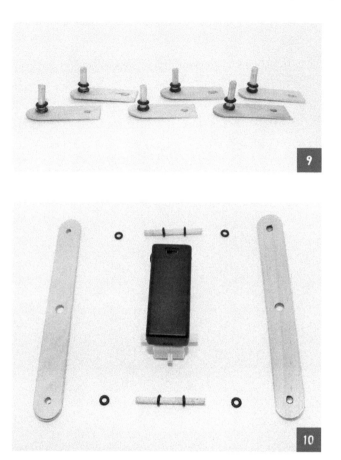

9. Once the glue is set, add another O-ring to each shaft of the six wooden rods.

PART B: ASSEMBLING THE CHASSIS (OR BODY)

10. Gather the motor and battery holder, the two 1¾-inch rods, eight O-rings, and the two double craft sticks with three holes. Place two O-rings on each shaft of the two wooden rods, about 1 inch apart.

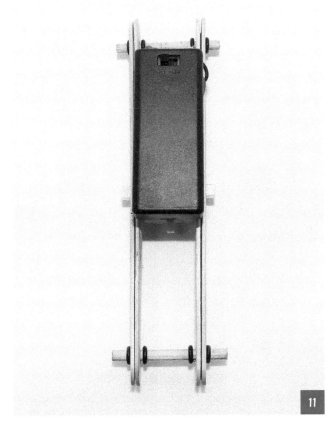

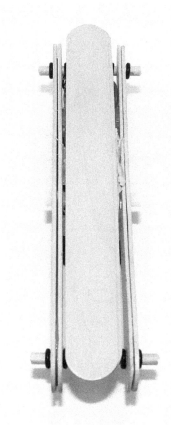

11. Insert the motor shafts through the center holes of the double craft sticks, and the wooden rods through each set of end holes. Place an O-ring on the ends of each wood shaft to secure the craft sticks. You may need to widen the center holes a bit with a hobby knife, depending on the size of the motor shafts.

12. Flip the **chassis** over. Glue a craft stick on the bottom of the motor for the base.

CONTINUED ➡

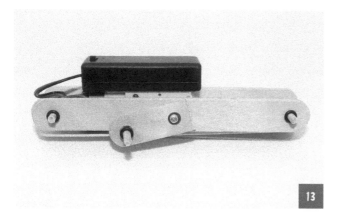

13

15

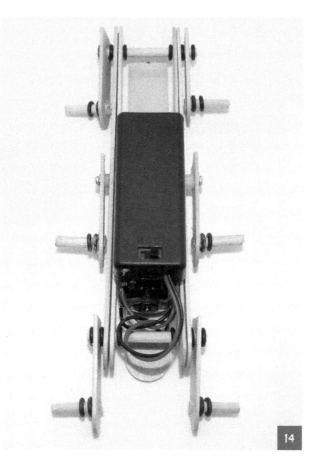

14

13. Using a small screw and washer, attach one of the craft stick pieces with a rod to one motor shaft. Screw in tightly. Repeat for the other side. Make sure both pieces line up evenly.

14. Attach the remaining four craft stick pieces to each of the wooden shafts at the ends of the chassis, and secure each one with an O-ring.

PART C: MAKING THE FEET

15. Glue the side of each quadruple craft stick to the top of a single craft stick. Make sure they line up along the inner edge, as shown. These are the feet.

16. Feed each shaft of the craft stick pieces through the appropriate hole on one of the feet. Make sure the flat part of the foot sticks outward. Secure each shaft with an O-ring.

17. Flip the robot upside down, and apply the magnetic tape to the base and to the bottoms of the feet. The base piece should be longer than the feet pieces. Flip the robot back over and turn on the motor. Try it on magnetic and nonmagnetic surfaces.

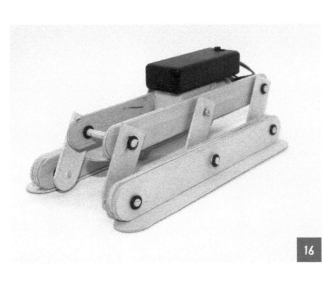

16

How it works: This robot alternates between stepping on its base and on its feet. Since it steps high, it can walk over uneven surfaces. When it walks up a magnetic surface, like a metal door, the magnet on its base holds it in place while the feet swing around for another step.

STEAM CONNECTION: Building, or manufacturing, each part is engineering. Measuring the distances of the holes uses mathematics. A battery-powered motor is a form of technology.

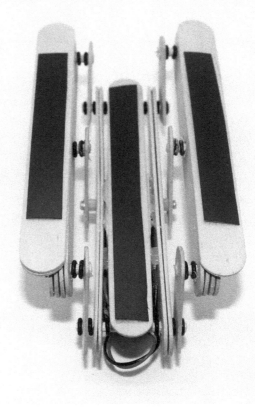

17

Chapter 6

ROBOTS ON THE JOB

When you look around your room, every-thing you see was thought up and built by a human. For a long time, people did all the work of not only thinking of ideas for new things but also of building them. But some work is dangerous. Some work is difficult. Some work is expensive. And some work is just plain boring. Sometimes, work can even be all those four things together. (Can you imagine?!)

People create robots to help humans do their jobs for a variety of reasons. Some-times, a robot will replace a human doing a specific job because a robot can do that job better, quicker, safer, and with more accu-racy. All of our computers and smartphones are built by robots in factories. Because machines always perform the same tasks in the same exact way, you can be sure that all computers and phones of the same brand

and model will work exactly the same. All the parts have been assembled with the same level of precision by a machine that cannot be distracted or become forgetful after working a long day.

So what are some other jobs that robots are doing today? Across many industries where products need to be assembled—from industrial bake shops to automobile factories—robots perform both simple, repetitive tasks, as well as tasks that require a high degree of concentration and skill from a human. Often, there is a human at the end of the process who performs quality checks and puts finishing touches on the product being created. However, some robots of industry work not just alongside but also collaboratively with humans. For example, Baxter and Sawyer, created by Rethink Robotics, are mobile warehouse helpers that can operate machines, do line loading, perform packaging, and more. In health care, RP VITA, by iRobot, allows various medical professionals and caregivers to gather in real time by a patient's bedside, regardless of where each person is actually located. RP VITA has autonomous navigation and mobility. RODIS is a robot designed to inspect oil pipelines to detect weak spots in the pipe structure, with arms to search for cracks. And Indago is a drone that uses infrared sensors to find wildfires and then tells airborne tankers exactly where to dump their water.

Many people are worried that robots will "steal" people's jobs. But only a human can put a robot in charge. A robot is just a tool. And if a robot can do a specific job better than a person, that frees up people to do other types of work that machines are not yet very good at doing, including using their imaginations to create even more advanced robot technology. When we push the limits of what our technology can achieve, we build on our STEAM **evolution**.

In this chapter, we will make some robots that can do a small fraction of some of the jobs we do all the time. We'll make one robot that cleans for us, one that moves around on water, and one that can sense obstacles.

BUILD A SWEEPING ROBOT

Cleaning up is *the worst*. If people didn't have to spend so much time cleaning, doing dishes, or washing ourselves, humans would probably be living on Mars by now. This next robot will sweep up dust and dirt for you so you can focus on what's really important: inventing a time machine.

In previous projects, we used small, disk-shaped vibration motors. This one uses a large vibration motor. If you can't find a large vibration motor, you can make one. A vibration motor is just a regular motor with an off-center weight on it.

TOTAL TIME: 20 MINUTES

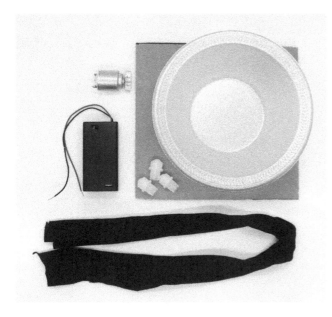

MATERIALS:

- 1 large vibration motor
- 1 (2-AA) battery holder with on/off switch and wire leads attached
- 2 AA batteries
- 1 small paper or Styrofoam bowl
- 1 square piece of cardboard, just bigger than the bowl
- 3 plastic ball casters
- 1 piece of fabric, a couple of inches wide and as long as the circumference of, or length around, the bowl
- Hot glue gun and glue stick
- Hobby knife or scissors
- Ruler
- Decorations like googly eyes and stickers (optional)

CAUTION: This project requires using a hot glue gun, which calls for extra caution so as not to get burned. Ask an adult for assistance. Also, ask an adult for help using the hobby knife if you're not completely comfortable using it on your own.

CONTINUED ➤

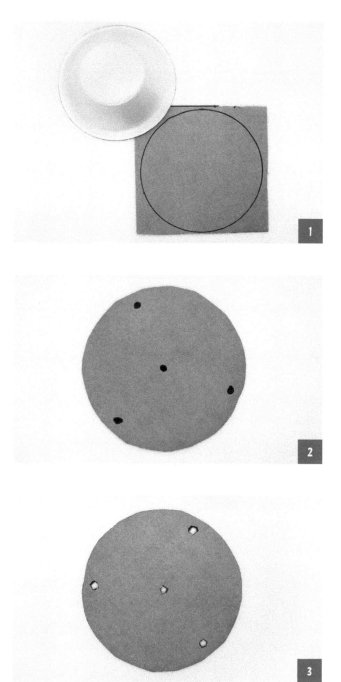

STEPS:

1. Trace the wide end of the bowl on the piece of cardboard. Cut the tracing out with scissors or a hobby knife.

2. Make a mark in the center of the cardboard circle. An easy trick to find the center of a circle is to use a ruler to find the widest part of the circle. That is the diameter. Measure that distance and divide it in half to find the middle. Make three marks about ½ inch from the edge of the circle, as equally apart as you can. It doesn't have to be exact, but get them as close to equal thirds as you can.

3. Cut out small holes on all the marks, about ¼ inch in width.

4. Insert the plastic ball casters in the outer holes.

5. Attach the wires from the battery holder to the motor.

6. Put a small ring of hot glue around the bottom edge of the motor. Very quickly place the motor in the center hole of the cardboard so the shaft is in the cutout.

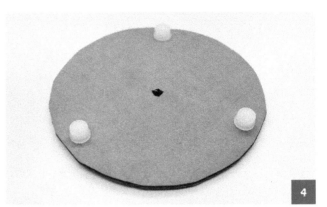

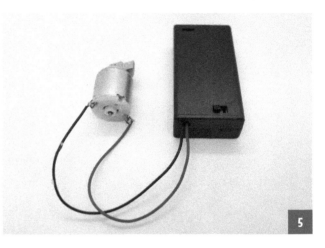

CONTINUED ➡

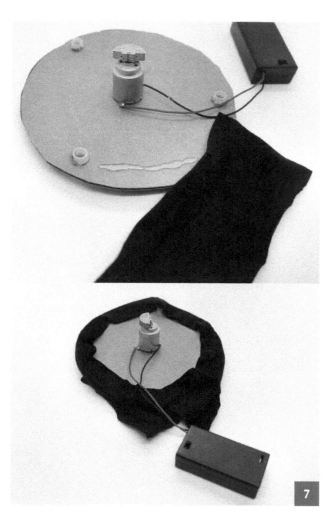

7. Glue one long edge of the fabric around the outside of the cardboard circle.

8. Glue the bowl, upside down, on top of the cardboard. Make a small slit for the battery holder wires so you can attach the battery holder on top of the bowl.

9. Cut slits in the fabric, about ½ inch apart, so the fabric pieces spread out to catch more dust. Decorate your sweeper robot however you like!

> **How it works:** The motion from the vibration motor spins the bot around on the casters. The fabric pieces pick up dirt and dust.

STEAM CONNECTION: You use mathematics when measuring with a ruler. Decorating the robot uses art. Manufacturing your own pieces for the different parts of your sweeper uses engineering.

BUILD A SOLAR BOAT BOT

Playing with boats is fun. This project will work great in a backyard swimming pool or in a nearby pond or creek. It is important to keep all our bodies of water clean. Discarded or lost batteries can be harmful to plants and animals that live in water, so we don't want to use any battery-powered motors. Without a battery, how will the boat bot move? Sailboats don't need motors, as they rely on the wind. You might be wondering, "What if there is no wind?" We'll create our own wind, using the same solar-powered circuit as the Solar-Powered Orbiter Mobile on page 57.

TOTAL TIME: 40 MINUTES

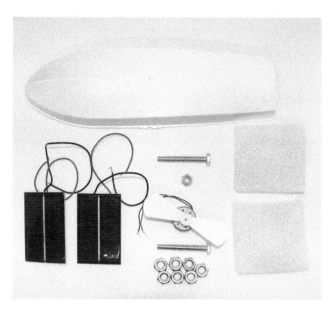

MATERIALS:

- 1 toy boat
- 2 small solar panels, rated 0.5V, 800mA, or higher
- 1 low-voltage, low-current motor with propeller attached (You can use the same motor and propeller from the orbiter mobile project.)
- 1 (¼-inch by 3-inch) bolt with lock nut
- 1 (¼-inch by 3-inch) bolt with 7 regular nuts
- 2 small squares of Styrofoam
- Hot glue gun and glue stick
- Hobby knife or scissors

! CAUTION: This project requires using a hot glue gun, which calls for extra caution so as not to get burned. Ask an adult for help. Also, ask for help using the hobby knife if you're not completely comfortable using it on your own.

STEPS:

1. Connect the red wire from one solar panel to the black wire of the other. (These solar panels are wired in series.) Connect the motor so that when the fan spins, it blows air away from it. Test the circuit in the sun, or under a very bright light.

2. Hand-tighten the lock nut to one of the bolts (right). This will be the motor mount. Screw all the regular washers onto the other bolt (left). This will be used for **ballast**, which is weight that balances a boat.

3. Glue the ballast bolt to the bottom inside of the boat, near the middle.

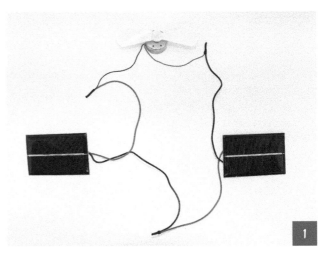

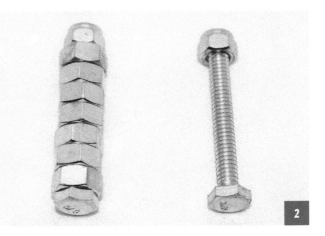

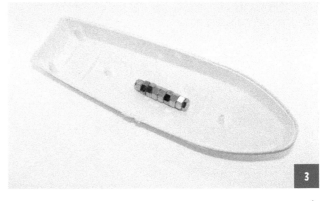

CONTINUED ➡

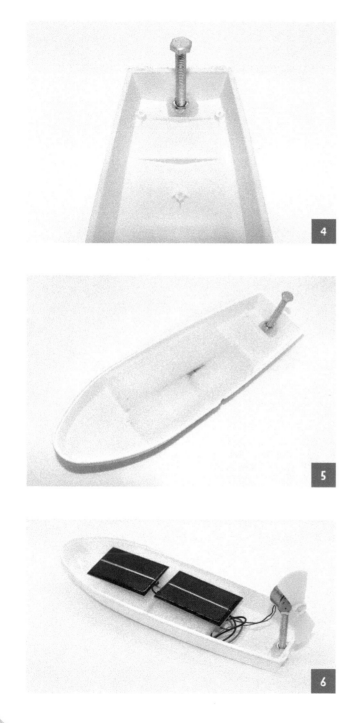

4. Glue the lock nut attached to the bolt to the bottom rear of the boat. Make sure you only glue the nut and not the bolt. This is the steering bolt.

5. Cut each square of Styrofoam in half to get four rectangular pieces. Place them on either side of the ballast bolt. Secure them with glue.

6. Place the solar panels over the Styrofoam pieces. Tuck the wires in between the Styrofoam. Glue the side of the motor to the top of the steering bolt, at a bit of an angle to the port (or left) side of the boat. Make sure the propeller does not hit the back of the boat as it spins.

7. Take your boat out into the sun, and the propeller should start spinning. Slightly twist the steering bolt to angle the propeller in different directions.

How it works: The solar panels turn light into electricity, powering the fan motor. You can angle the fan to control the direction of the boat.

STEAM CONNECTION: Building a solar-powered boat is a form of technology. Finding the right balance so the boat doesn't tip requires engineering skills.

BUILD A CLIFF-DETECTING ROBOT

The Roomba vacuum has many **cliff sensors**, which tell it if it's about to fall down the stairs or off an edge. Its cliff sensors are infrared LEDs. The next robot we will build also has a cliff sensor, but it is mechanical instead of electrical. Once the robot gets to an edge, it spins out of the way.

TOTAL TIME: 1 HOUR AND 15 MINUTES

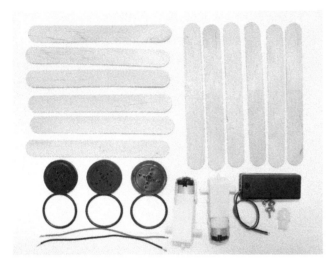

! CAUTION: Be careful when twisting the wires onto the motor. If an adult has a **soldering iron**, ask if they can solder the wires for you. If using a hot glue gun, use extra caution so as not to get burned. Ask an adult for help. Also, ask an adult for help if you're not comfortable using the hobby knife on your own.

MATERIALS:

- 12 (6-inch) tongue depressors
- 1 (2-AA) battery holder with on/off switch and wire leads attached
- 2 AA batteries
- 1 (6-inch) piece of red wire, stripped on both ends
- 1 (6-inch) piece of black wire, stripped on both ends
- 1 high-speed gearbox
- 1 low-speed gearbox
- 3 wheels (best if 2 fit the low-speed gearbox and 1 fits the high-speed gearbox)
- 3 medium rubber gaskets for tires
- 3 small screws
- 1 plastic ball caster
- Hot glue gun and glue stick
- Hobby knife or scissors
- Pencil

CONTINUED ➡

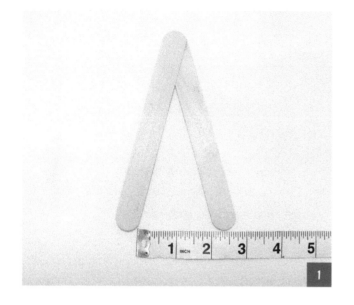

STEPS:

1. Hot-glue the ends of two tongue depressors together so the inside edges are 2 inches apart.

2. Glue another tongue depressor to one of the two, starting about a quarter of the way down so the bottom of the third depressor extends beyond the one it's glued to.

3. Flip the glued "V" over, and repeat the previous step on the other side.

4. Stretch the gaskets onto the wheels like tires. Attach two wheels to the low-speed gearbox and one wheel to the high-speed gearbox. Fasten them with small screws if you can, or with a dab of glue.

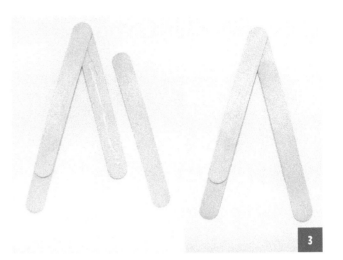

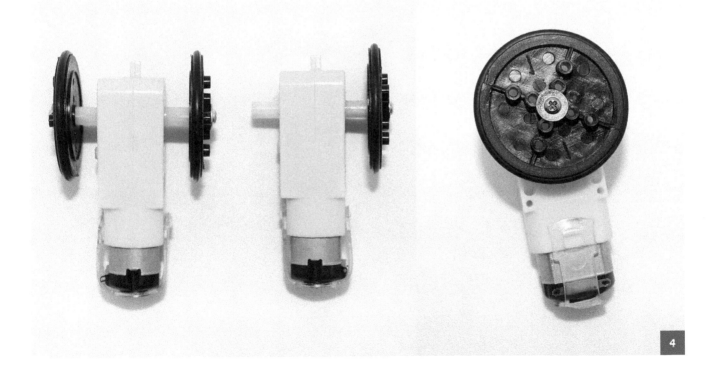

CONTINUED ➡

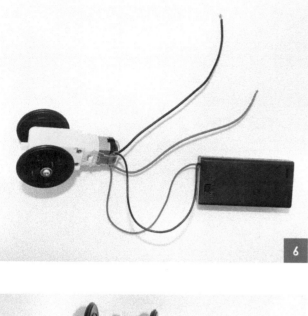

5. Twist the end of the 6-inch piece of red wire to the end of the red wire on the battery holder. Repeat this step with the black wires.

6. Attach the twisted ends of the wires to the low-speed gearbox with two wheels, so it moves forward with the wheels at the back.

7. Attach the remaining wire ends to the single-wheel gear motor. The two motors are now connected in parallel.

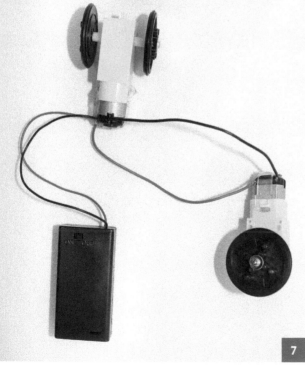

8. Glue the two-wheeled motor to the back of the "V," making sure the wheels can spin freely.

9. Glue the second motor perpendicular, or at a right angle, to the first. Make sure the single wheel is as close as possible to the inside of the "V" without touching it.

10. Add another tongue depressor as close to the third wheel as possible without touching it, changing the "V" to an "A."

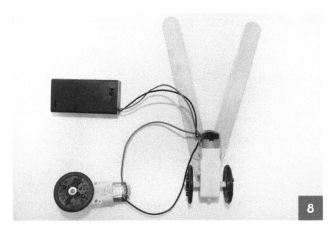

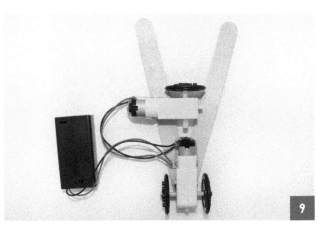

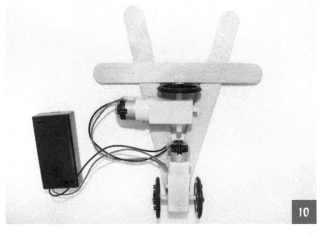

CONTINUED ➜

11. Glue on a small end piece from another tongue depressor to make the left side even with the right side. Measure each leg to see exactly how much shorter the short leg is, and then use that measurement to cut your small end piece.

12. Place the ball caster below the horizontal line of your wooden "A," at an equal distance from both legs. Press a tongue depressor under it to hold the plastic ball caster in place and mark the spot where the unglued tongue depressor sits with a pencil.

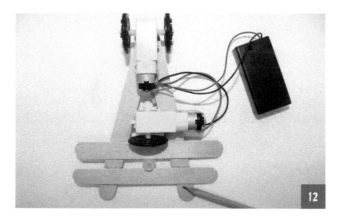

13. Glue the tongue depressor in the spot you marked in the previous step.

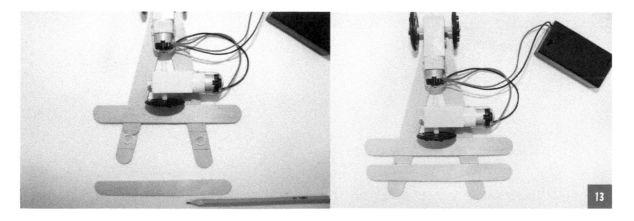

14. Once the hot glue has set, press the plastic ball caster in between the cross braces.

15. Next, cut three tongue depressors in half. Glue them all together in a stack, with the rounded ends facing the same way.

16. Glue the stack on top of the two-wheeled motor. Then, glue the battery holder, switch side up, on top. The stack of cut tongue depressors is essentially a base and riser for your battery holder.

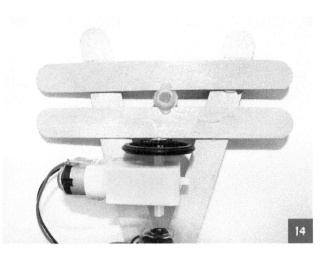

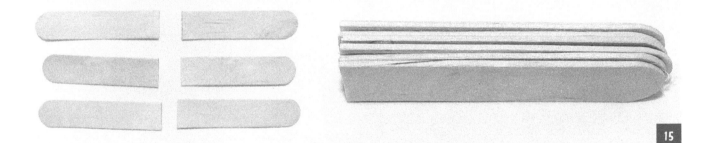

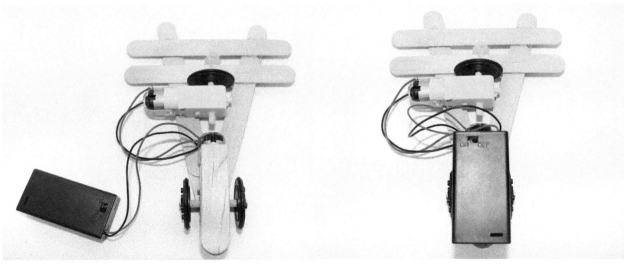

CONTINUED ➡

17

17. Adjust the height of the roller ball so it sits just a little lower than the wheel next to it.

18. Cut off the rounded ends of your tongue depressor structure to make it shaped like a pizza slice. Turn on your robot, set it on a raised surface like a tabletop, and observe.

How it works: The robot is held up by the two rear wheels and the front caster. When the front drives off an edge of, say, a table, the third wheel engages and turns the robot sideways before it can roll off the table.

STEAM CONNECTION: Figuring out how different parts go together to make a robot work, and troubleshooting to get a desired result, is engineering. Measuring parts involves mathematics.

Chapter 7

ROBOTS IN THE OPERATING ROOM

Humans have advanced medicine in amazing ways. But there are things humans cannot do. Robots can be made smaller, stronger, and more precise than human hands. Robots that perform surgery fit tools and cameras into tiny openings. Because it takes patients less time to heal from small cuts, the use of robots in the operating room has improved many procedures, allowing people to check into a hospital one day and be back at work and play within days of what would have been considered major surgery in the past.

Robotic surgery does not replace surgeons. Usually, a surgeon is in the room, controlling the robot from a terminal, while medical staff are nearby, ready to step in when they're needed. The surgeon controlling the robot communicates with the other medical staff on standby using an audio system. The da Vinci surgical robots, which come equipped with advanced and tiny instruments, provide surgeons a high-definition and three-dimensional view of the surgery area. These robots have performed millions of surgeries, according to Intuitive Surgical Inc., the company that makes them. They are used for all kinds of delicate surgeries, including those on the heart and spine, with fewer post-surgical complications than operations done by human hands.

Think about it. Surgeons often work very long hours doing tasks that require a lot of mental focus and steady hands. Operating a computer console, with a high-definition view of the surgical area, takes less out of surgeons than traditional surgeries. Less mental, eye, and physical (hand) fatigue, or tiredness, means that those surgeons can perform more surgeries better. Plus, tiny robotic arms and instruments are able to access hard-to-reach parts better than a surgeon's hands, meaning there's less risk of mistakes being made, and fewer complications for patients to heal from.

Another surgical robot that is revolution-izing the world is called the CorPath System. This robot uses virtual reality to allow surgeons to practice complex procedures before actually doing them. Plus, surgeons can operate the robot remotely, so a doctor in one part of the world can perform a surgery on a patient on the other side of the world. Many people around the world live in remote areas. This robot allows such people access to lifesaving surgeries, even when there are no surgeons nearby. It also allows very sick patients who cannot be moved to get the surgeries they need in whatever hospital they happen to be in.

Despite these amazing advances, robots will never completely replace humans in the operating room because surgery can be unpredictable and human surgeons and medical staff must always be ready to jump in if something goes wrong during a proce-dure. But, with robotic help, doctors can heal people faster.

For this chapter, we will build robots with different forms of **locomotion**, or movement, that a surgical robot might need to perform in an operating room. Although surgical robots are extremely complex, their engineers had to begin by considering the kinds of movements that they would need to be able to do.

BUILD A FOUR-LEGGED WALKING ROBOT

Different robots use different forms of locomotion. Some robots have wheels. Some robots have propellers. Some robots have jet engines. And some robots have legs, like humans or animals. The next few projects in this chapter will focus on different forms of locomotion. We will use the same gearbox, as well as many of the same parts, for the next three robots.

The first robot we will build has four legs. This is one of the most stable robots. When we first learn to move, we crawl on all fours. It is easier to get around on all fours without losing your balance.

TOTAL TIME: 1 HOUR AND 30 MINUTES

NOTE: Many of these depressors can be reused for some of the next projects.

MATERIALS:

- 1 Elenco 2-in-1 gearbox kit
- 1 (2-AA) battery holder with on/off switch and wire leads attached
- 2 AA batteries
- 12 (6-inch) tongue depressors
- 1 (4-inch) wooden rod, ⅛ to ³⁄₁₆ inch in diameter
- 1 (2-inch) wooden rod, ⅛ to ³⁄₁₆ inch in diameter
- 2 small screws
- 4 small cotter pins
- 10 small rubber O-ring gaskets, ⅛ inch in inner diameter (must fit snugly on wooden rods)
- Hot glue gun and glue stick
- Hobby knife
- Drill with two different-size drill bits
- Scissors
- Pencil or marker

CONTINUED ➡

! **CAUTION:** Ask an adult for help with the hot glue gun, hobby knife, and drill. When drilling the holes, the large hole needs to be exactly the size of the wooden rods. The small hole needs to be able to let the "legs" of the cotter pins through, but not the head.

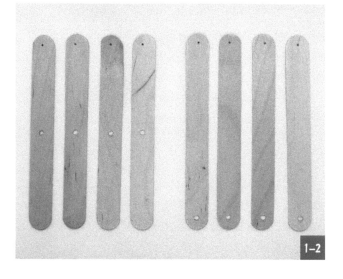

1–2

STEPS:

PART A: BUILDING THE LEGS

1. Take four tongue depressors and ask an adult to help you drill one hole the size of the wooden rod into each of the depressors, exactly in the middle. Then, ask the adult to drill a small hole for a cotter pin about ¼ inch from one edge of the depressors. These will be the legs.

2. Take four more tongue depressors and have the adult drill a hole the size of the wooden rod, about ¼ inch from one end, into each of them. Then, have the adult drill a tiny hole for the cotter pin about ¼ inch from the other end of each of the depressors. These will support the legs. Your adult should have drilled a total of 16 holes.

3. Connect the tongue depressors by type so you have pairs, connecting one "leg" with one "supporter," at the ends with the smaller holes. Insert the thin end of the cotter pin until its head is stopped by the hole. Then, bend the ends away from each other to hold the two pieces together.

PART B: BUILDING THE CHASSIS (OR BODY)

4. Glue two of the remaining tongue depressors together. Repeat this step so you have a pair of double tongue depressors. This will be for the chassis. Line both double tongue depressors up to the metal gearbox case, and mark where the holes are for both sides.

5. On the inner mark of each depressor, trace around one of the blue **bushings** from the gearbox kit.

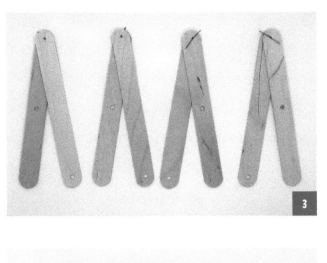

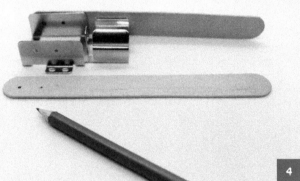

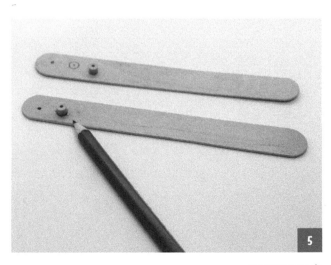

CONTINUED ➡

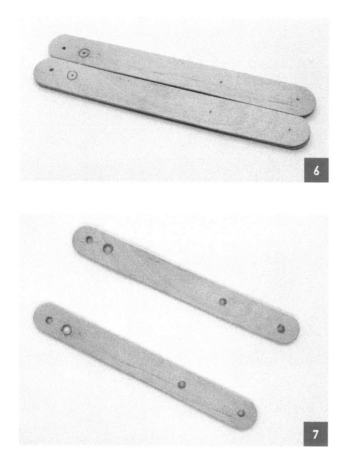

6. Make other marks about ½ inch from the opposite edge, then about 2 inches from that same edge. There should be a total of eight marks, four on each depressor.

7. Have an adult drill all four marks with the larger drill bit, the size of the wooden rod. Widen out the hole for the blue bushing.

8. Build the gearbox using the enclosed instructions and the slow gear parts with the long shaft. Do not put the white ends on the shaft yet.

9. Place one of the double tongue depressor pieces (with four holes) on each side of the gearbox, inserting the long shaft through the small holes nearest the wider holes. Make sure the blue bushings fit in those wider holes. Slide the 4-inch rod through the two holes near the other end. Put the shorter rod in the holes near the middle. Secure the rods with six gaskets as shown.

10. Press the white **crank arms** onto the metal gearbox shaft. They should face in opposite directions.

11. Add an additional gasket near each end of the longer wooden rod, which will be used to hold the legs in place.

12. Flip the whole base over, so the flat side of the gearbox is facing up. This will help protect the gears and your fingers. Twist the red wire from the battery case through the hole in the metal terminal marked (+) on the motor. Twist the black wire to the other metal terminal.

13. Place the battery holder behind the gearbox, over the two wooden rods, with the switch facing up. You can hold the battery holder in place with a dab of glue or some tape.

PART C: ATTACHING THE LEGS

14. Form two sets of the legs and supports into an "M," so that the hole at the end of one tongue depressor pair lines up with the middle hole of the other.

15. Screw the left-hand holes into the white crank arm at the end of the metal axle. Remember to ask an adult anytime you want to use power tools.

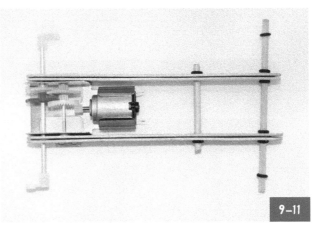

9–11

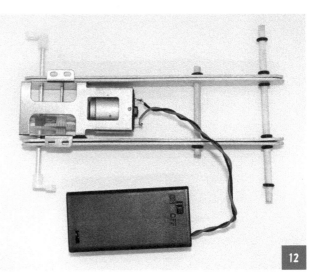

12

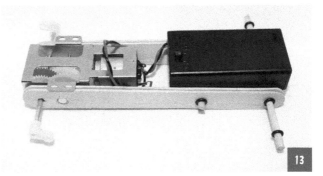

13

CONTINUED ▶

16

16. Slide the right-hand holes onto the two wooden rods. Use a gasket on each axle to hold the "M" in place.

17. Flip your robot around. Just like you did with the other side, form two sets of the legs and supports into an "M," so that the hole at the end of one tongue depressor pair lines up with the middle hole of the other.

18. Now, screw the right-hand holes into the white crank arm at the end of the metal axle. This time, slide the left-hand holes onto the wooden rods. Use gaskets as before to hold the second "M" in place.

19. Put batteries in your battery holder and turn the switch on. Your robot should start walking around.

> **How it works:** The drive shaft for each leg is **offset**, so when one is up, the other is down, just like when we take steps. Each step of the front leg controls the back leg on the same side.

STEAM CONNECTION: Fabricating, or making your own parts, is engineering. Measuring distances requires the use of mathematics. Exploring how four-legged animals walk is science, and turning those observations into a robot is technology. If you decorate your robot, that is art.

19

BUILD A HEXAPOD ROBOT

The next robot we will build has six legs, which is called a **hexapod**. All insects have six legs. Having six legs helps them walk over uneven terrain. Because insects are so small, ground that looks flat to us might look like a mountain range to them. Six legs help them stay balanced as they move around their bumpy surroundings.

TOTAL TIME: 2 HOURS

CAUTION: Ask an adult for help with the hot glue gun, hobby knife, and drill. When drilling the holes, the large hole needs to be exactly the size of the wooden rods. The small hole needs to be able to let the "legs" of the cotter pins through, but not the head.

MATERIALS:

- 1 Elenco 2-in-1 gearbox kit
- 1 (2-AA) battery holder with on/off switch and wire leads attached
- 2 AA batteries
- 20 (6-inch) tongue depressors
- 2 (4-inch) wooden rods, ⅛ to ³⁄₁₆ inch in diameter
- 2 small screws
- 6 small cotter pins
- 16 small rubber O-ring gaskets, ⅛ inch in inner diameter (must fit snugly on wooden rods)
- Hot glue gun and glue stick
- Hobby knife
- Drill with two different-size drill bits
- Scissors

NOTE: You can reuse some of the leg and support pairs from the last project. This project uses 6 "leg" tongue depressors (with large middle hole and small end hole) and 6 "support" tongue depressors (with large and small end holes).

STEPS:

PART A: BUILDING THE LEGS

1. Take six tongue depressors and have an adult help you drill a hole the size of the wooden rod into each of the depressors, exactly in the middle. Drill a small hole for a cotter pin about ¼ inch from one edge into each of the depressors. These will be the legs.

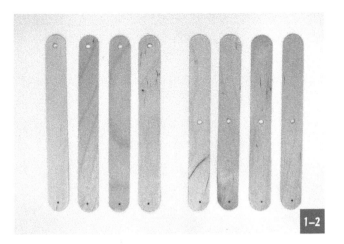

2. Take six tongue depressors and have your adult drill a hole the size of the wooden rod, about ¼ inch from one end, into each of them. Drill a tiny hole for the cotter pin about ¼ inch from the other end into each of the depressors. These will support the legs. You should have 24 holes in all.

3. Connect the tongue depressors by type so that you have pairs, connecting one "leg" with one "supporter." Connect them through their small holes using cotter pins by inserting the thin end of the cotter pin until its head is stopped by the hole. Then, bend the ends away from each other to hold the two pieces together, like a hinge.

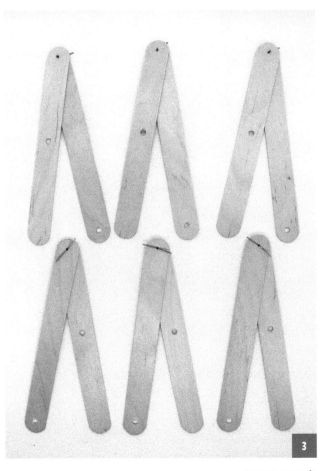

CONTINUED ➡

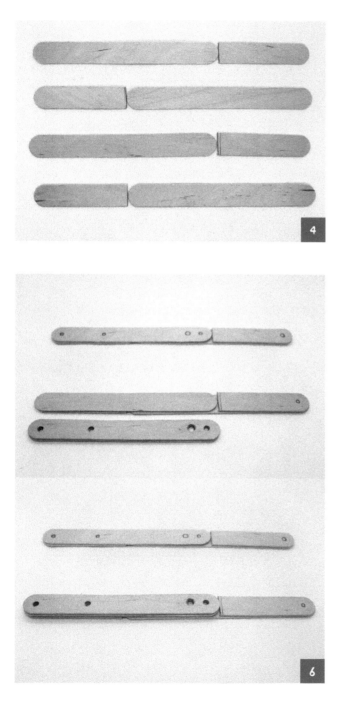

PART B: PREPARING THE TONGUE DEPRESSORS FOR THE BODY

4. Cut two of the undrilled tongue depressors in half, using a sharp hobby knife or scissors. Ask an adult for help if using a knife. Arrange four rows of the full and half tongue depressors in a brick pattern as shown.

5. Glue the top two, and bottom two, rows together. You will have two pieces that are 12 inches long, or double-length, double-thickness tongue depressors.

6. You can use a main body tongue depressor from the four-legged robot to mark the holes for the left half as shown. The right-most hole should be about 1¼ inches from the edge.

7. Ask an adult to help you drill all ten marks with the larger drill bit, the size of the wooden rod. Widen out the hole for the blue bushing of the gearbox.

PART C: MAKING THE GEARBOX

8. Build the gearbox using the enclosed instructions and the slow gear parts, with the long shaft. Do not put the white crank arms on the shaft yet. If you are reusing the gearbox from the four-legged robot, you can leave the battery wires attached.

PART D: ADDING THE POWER SOURCE

9. Twist the red wire from the battery case through the hole in the metal terminal marked (+) on the motor. Twist the black wire to the other metal terminal. Flip the gearbox over, so the flat side is up. Carefully glue or use double-sided tape to stick the battery holder (switch side up) to the flat side of the gearbox.

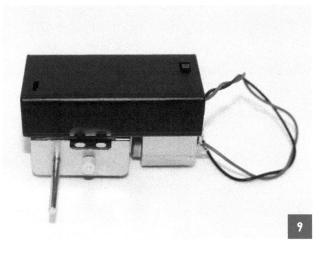

PART E: BUILDING THE CHASSIS (OR BODY)

10. Place one of the long body pieces (with four holes) on each side of the gearbox, inserting the long metal shaft through the center holes. Make sure the blue bushing fits in the widest hole. Slide a 4-inch rod through the two holes near each end. Secure the rods with eight gaskets as shown.

11. Press the white crank arms onto the metal axle, facing in opposite directions. Add an extra gasket near each end of the two wooden rods, which will hold the legs in place.

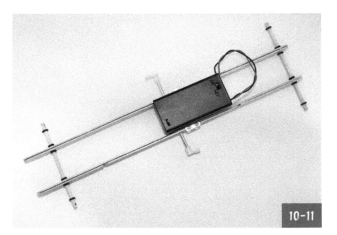

PART F: ATTACHING THE LEGS

12. Align the body so the wooden axle farthest from the edge is on the left. Slide the middle hole of one of the leg pieces onto the right-most wooden axle. Use a gasket to hold it in place.

CONTINUED ➡

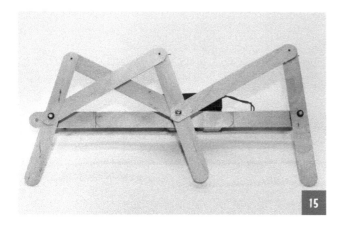

13. Form two sets of the legs and supports into an "M," so that the hole at the end of one tongue depressor pair lines up with the middle hole of the other.

14. Slide the left-hand holes onto the leftmost wooden axle. Use a gasket to hold it in place.

15. Line up the right-hand holes with the end hole of the remaining leg support. Place a screw through all three pieces and attach them to the white crank arm.

16. Spin the robot around, and repeat the same steps for the legs, but in a **mirror image**. Slide the middle hole of one of the leg pieces onto the leftmost wooden axle. Use a gasket to hold it in place.

17. Form two sets of the legs and supports into an "M," so that the hole at the end of one tongue depressor pair lines up with the middle hole of the other.

18. Slide the right-hand holes onto the right-most wooden axle. Use a gasket to hold it in place.

19. Line up the left-hand holes with the end hole of the remaining leg support. Place a screw through all three pieces and attach them to the white crank arm.

20. Put batteries in your battery holder and turn the switch on. Your hexapod robot should start walking around.

> **How it works:** The middle leg on each side controls the front and back legs. This allows you to use a single motor to control all six legs.

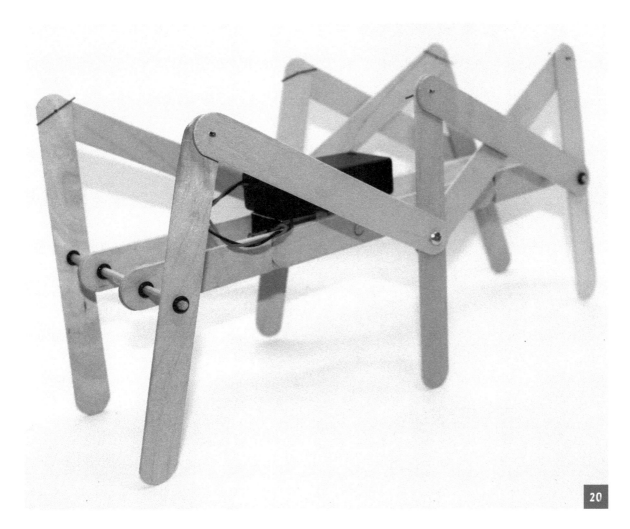

20

BUILD A BIPEDAL ROBOT

The last robot we will build is a **bipedal**, or two-legged, robot. When humans or animals walk around on two legs, we must balance on one leg during each step. Our ankles move side to side to help us move our center of gravity to stay balanced. We also have fluid in our ear canals that helps us tell if we are unbalanced, or about to fall over. Keeping bipedal robots balanced is challenging. Some humanoid robots have servos for "ankles" to help them balance. They may also have **tilt sensors** or even **gyroscopes** to help keep them balanced. Our bipedal robot will have two **trusses**, or braces, to help each leg hold up the robot while it takes a step.

TOTAL TIME: 2 HOURS AND 30 MINUTES

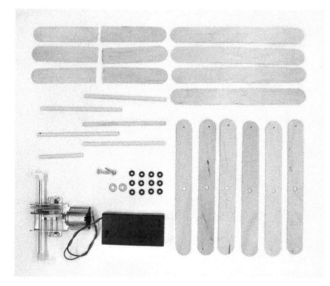

NOTE: You can reuse some of the leg pieces from the four-legged robot project on page 109. This project uses four "leg" tongue depressors (with large middle hole and small end hole).

MATERIALS:

- 1 Elenco 2-in-1 gearbox kit
- 1 (2-AA) battery holder with on/off switch and wire leads attached
- 2 AA batteries
- 13 (6-inch) tongue depressors
- 5 (4-inch) wooden rods, ⅛ to 3⁄16 inch in diameter
- 1 (2-inch) wooden rod, ⅛ to 3⁄16 inch in diameter
- 2 small screws with washers
- 2 small cotter pins
- 12 small rubber O-ring gaskets, ⅛ inch in inner diameter (must fit snugly on wooden rod)
- Hot glue gun and glue stick
- Hobby knife
- Drill with two different-size drill bits
- Scissors

CAUTION: Ask an adult for help with the hot glue gun, hobby knife, and drill. When drilling the holes, the large hole needs to be exactly the size of the wooden rods. The small hole needs to be able to let the "legs" of the cotter pins through, but not the head.

STEPS:

PART A: PREPARING THE TONGUE DEPRESSORS

1. Cut four of the tongue depressors exactly in half with the scissors, ending up with eight pieces.

2. Take four tongue depressors and ask an adult to drill a ⅛-inch hole, the size of the wooden rod, exactly in the middle of each of them. Drill the small hole for the cotter pin about ¼ inch from one edge in each of them.

PART B: PREPARING THE BODY

3. Glue two of the undrilled tongue depressors together. Repeat, so you have two double tongue depressors. These will be for the robot chassis.

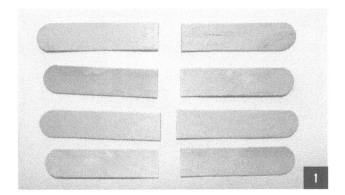

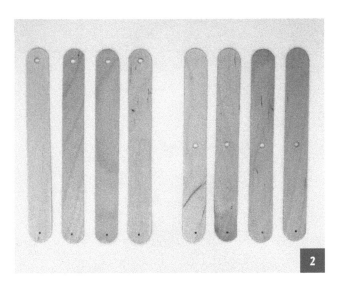

CONTINUED ➡

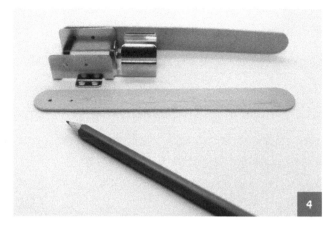

4. Line both double tongue depressors up to the metal gearbox case, and mark where the holes are for both sides. On the inside mark, trace around one of the blue bushings from the gearbox kit.

5. Make another mark about ½ inch from the opposite edge, then another about 2 inches from that same edge. There should be four marks in total per piece.

6. Ask an adult to help you drill all four marks with the larger drill bit, then widen out the hole for the blue bushing.

PART C: MAKING THE GEARBOX

7. Build the gearbox using the enclosed instructions and the slow gear parts, with the long shaft. Do not put the white crank arms on the shaft yet.

8. Twist the black wire from the battery case through the hole in the metal terminal marked (+) on the motor. Twist the red wire to the other metal terminal.

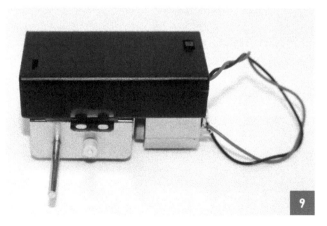

9. Stick the battery holder to the flat bottom of the gearbox using hot glue or double-sided tape. Make sure the switch is sticking up.

PART D: BUILDING THE CHASSIS (OR BODY)

10. Connect one of the double tongue depressor pieces (with four holes) to each side of the gearbox, inserting the long shaft through the small holes nearest the wider holes. Make sure the blue bushing fits in the wider holes.

11. Put one of the 4-inch wooden rods through the holes closest to the gearbox. Hold it in place with four gaskets.

12. Place the 2-inch wooden rod through the front holes. Hold it in place with four gaskets.

13. Press the two white crank arms from the gearbox kit onto the ends of the metal shaft, facing them in opposite directions.

PART E: BUILDING THE LEGS

14. Start with six of the half tongue depressor pieces. Glue two pairs of them together. This will leave two single halves.

15. Stack all the half pieces together and ask an adult to help you drill a hole (sized for the wooden rods) near both bottom corners of one side. If you stack them and then drill, all the holes should line up.

16. Get four "leg" tongue depressors, with the large middle hole and small end hole. Glue two pairs together, so each leg is double strength.

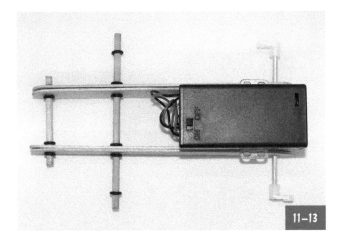

11–13

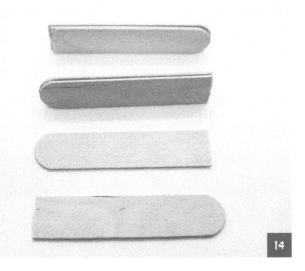

14

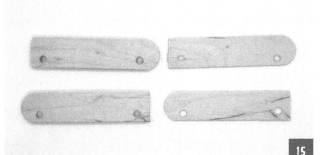

15

CONTINUED ➡

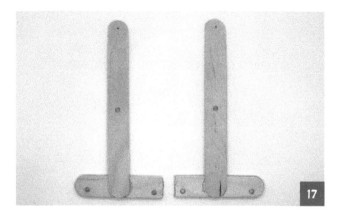

17. Glue the end of each double leg piece, without a hole, in the middle of each double half piece, with the holes at the bottom, as shown. Make sure the bottom half pieces are facing in opposite directions.

18. Insert the remaining 4-inch rods into the bottom holes, so the legs stand up.

19. Glue the two drilled single half pieces on the back of the legs. Make sure the wooden rods are not sticking out the back of the bottom half pieces too far, and line up the rounded ends with the double half pieces.

20. Drill a small and a large hole in the last two half pieces, no more than 2 inches apart, as shown.

21. Attach these two half pieces to the top of the legs by pushing the skinny end of the cotter pins through the small holes, then bending both sides out.

PART F: CONNECT THE LEGS TO THE BODY

22. Use a screw and washer to attach each leg to the white plastic crank arms on the metal shaft.

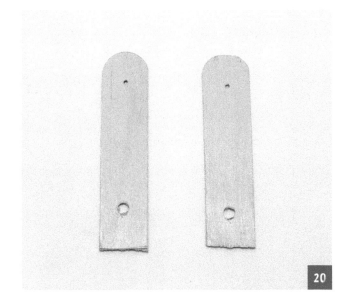

CONTINUED ➡

23. Use four gaskets to secure the two half tongue depressor pieces to the longer wooden rod on the chassis.

24. Put batteries in the battery holder and turn the switch on. Your bipedal robot should start walking around. If it falls over forward, try making the holes on the support arms closer than 2 inches apart to change the robot's center of gravity, making the body more upright.

25. Don't forget to decorate your robot with extra parts, markers, stickers, or whatever you can find. Make it look like a cool person, animal, or dinosaur.

> **How it works:** The pegs sticking inward on each leg help balance the robot, since it doesn't have any ankles, like a person does.

STEAM CONNECTION: Fabricating, or making your own robot parts, is engineering. Measuring distances requires using your mathematics skills. Exploring how two-legged robots balance is science. Turning those observations into a self-balancing robot is technology. If you decorate your robot, that is art.

Chapter 8

ROBOTICS AND THE FUTURE

Robotics is not only cool and *super* fun, but it's really important for the future of humanity. Robots help us move beyond our bodies to accomplish great things. Our human bodies are wonderful, but there is a lot they cannot do. The advantage of robots is that we can design them to perfectly accomplish the tasks we want them to do.

Humans may have walked on the moon, but space exploration is dominated by robots. Robots have visited all the planets in the solar system, and two robots, Voyager 1 and Voyager 2, have even left our solar system to explore beyond our sun. Robots are really helpful to humans in space, scouting safe places for humans to land and, eventually, sending supplies to distant planets to help humans survive away from Earth.

Although people may still have to wait a while for their own personal robot assistant, there are some exciting robots on the horizon. One of the challenges of having robot assistants is getting them to walk steadily on two legs, without falling over. Boston Dynamics' two-legged robot, Atlas, can walk on snow, jog, jump over obstacles, and even do backflips.

One of the most important innovations in robotics might not be in their bodies, but in their brains. **Artificial intelligence** (AI) is the ability of a computer to learn on its own. Instead of being programmed to do a task, AI robots can *learn* to do it on their own. This can be very helpful, but it also scares some people, who worry that AI robots may one day try to take over the world. But here's an important thing to remember: If you know how to build robots, then you know how to take them apart.

If you love robots and want to have a career in robotics, there are many great options you could pursue. You could become a robotics engineer and oversee the creation of robots. You could become a robot technician and fix robots when they break. Or you could become a software engineer who writes the code that runs robot brains. Since robotics involves every aspect of STEAM, there is almost no limit to the different paths you can take to spend your life working with robots—just like I do!

Resources

There are many great resources for people to learn about robots. Following are a few you can check out to continue your robotics education:

- https://robots.ieee.org/
  Information on all different types of robots.

- https://meetedison.com/
  A cool, LEGO-compatible robot for learning different programming languages.

- https://learn.adafruit.com/
  A great site for learning to make do-it-yourself projects.

- https://hackaday.com/
  A funny blog about hacking everyday objects or making weird things.

- https://makezine.com/
  A website dedicated to do-it-yourself/maker culture.

- https://www-robotics.jpl.nasa.gov/
  Information about NASA's robots.

- https://www.vexrobotics.com/
  The website of one of the largest robotics competitions in the world.

- https://www.bostondynamics.com/
  The homepage of one of the world leaders in robotics.

- https://www.khanacademy.org/
  A place for free education on all subjects.

- https://scratch.mit.edu/
  A free programming language to get you started coding.

Glossary

acronym: A word made up of the first letter, or letters, of other words that express an idea, such as STEAM (science, technology, engineering, art, and mathematics)

actuator: A part of a robot that moves something else, such as motors, gears, pulleys, and servos

android: A type of robot that looks or acts like a human, such as C-3PO, the Iron Giant, and WALL-E

art: The expression of creativity in a visual format

artificial intelligence: A computer program that can learn on its own, without human input

automaton: A self-moving machine that uses gears and levers to create actions, such as playing instruments

axle: A rod or shaft that goes through the center of gears or wheels upon which they spin

ballast: Something used as a weight at the bottom of a boat to keep it balanced in the water

ball caster: A sphere, or ball, in a holder that can support weight and can move freely in all directions, like the trackball in a traditional computer mouse

barometer: A type of weather instrument that measures the pressure of the air around it, used to predict the weather by comparing different pressures over time

battery: A power source that uses a chemical reaction to provide energy to electric circuits

bipedal: Having two legs for walking

breadboard: A base to plug electronic parts into, used for prototyping electric circuits

bump sensor: A mechanical lever, switch, or device that sends a signal to a robot's controller when it touches something

bulb receptacle: An enclosure that holds a light bulb

bushing: A piece of rubber or plastic that helps stop vibrations in a machine

capacitor: An electrical component that stores electrical energy, kind of like a battery

center of gravity: The point or place in any object where it is completely balanced

chassis: The base structure of a robot, car, or machine

circumference: The outside border of or distance around any shape, often a circle

civilization: The way different groups of people organize their lives for mutual betterment

cliff sensor: A device used to detect the edge of a surface

computer: A complex controller that can perform many input and output operations per second

conductor: The part of a material that electricity can flow through

controller: The "brain" of a robot, used to turn sensor inputs into movement

cotter pin: A metal piece of hardware used to connect two things together through a hole or small opening, with ends that can be flared or bent back

crank arm: A lever at the end of an axle that helps turn the axle's spinning motion into an up-and-down or back-and-forth, motion

decode: To try to understand how a computer's system or program works

design: To plan how a robot will look and act before building it

device: An object, usually electrical or mechanical, made to carry out a specific function

diameter: The longest distance across a shape, usually a circle

effector: The part of a robot that touches its surroundings, usually attached to an actuator, such as wheels, legs, and arms

electric circuit: A circular or looped path through which electricity can flow

electricity: A type of energy made up of charged particles, like electrons, that can be delivered through a current, or flow of charge

electric signal: A pulse of electric current that sends information from one component to another

electrolytic: Containing an electrolyte, a non-metallic electric conductor

electron: A negatively charged particle that orbits the center of an atom

empowering: Providing a feeling of confidence, for example, after you accomplish something difficult and realize you can do something you didn't know you could do, like building robots

engineering: Using science and mathematics to build robots, machines, buildings, roads, bridges, and other structures that are useful to people

evolution: The process by which people or things slowly change and improve over time

ferromagnetic: Being made of materials that are highly magnetic or easily magnetized

found object: Something used in art or robotics differently than it was originally intended to be used

gasket: A ring of flexible material, usually rubber, used to seal or hold two other materials together

gear: A wheel with equal cuts around it that make little spikes, called teeth

gearbox: Many gears connected to each other and mounted in a container

gyroscope: A spinning wheel that can freely tilt in different directions, so it can keep its original direction of spin

hexapod: A robot or animal with six legs

homopolar motor: A simple circuit that uses a battery, magnet, and wire to create rotational, or spinning, movement

humanoid: Having human qualities

hydraulics: A type of technology that uses liquids pushed or sucked through a tube or pipe to move something heavy

infrared LED: A type of light-emitting diode that shines reddish in color but is just beyond what can be seen by human eyes, used for communicating with remote controls

innovation: The act of creating new things, like inventions, robots, or products, that help people

input: Information, or data, from a sensor that is used by a computer or machine to act in a certain way

insulator: A type of material, like rubber or plastic, through which electricity cannot flow

International Space Station (ISS): A human-made structure orbiting Earth in which astronauts can live and perform science experiments outside our atmosphere and gravity

lever: A simple machine that uses a bar on a point, or fulcrum, to move a heavy load, such as the beam of a seesaw with the fulcrum at the middle

light-emitting diode (LED): A type of semi-conductor that produces photons, or light, as electric current passes through it

locomotion: The action of moving from one place to another

low-voltage, low-current motor: A type of motor that can spin using very little voltage and current, often powered by a solar panel

mathematics: The science of how numbers relate to one another, and also of how they describe our universe

mechanical: Produced by a machine

mechanics: The study of how different objects in motion can have an effect on one another

microcontroller: A type of small computer that can perform simple tasks

microphone: A transducer, or a device that transfers one type of input into a different output, that turns sound into an electric signal

mirror image: A reflection across an axis, or middle line

mobile: A type of kinetic, or moving, sculpture that balances and spins around, sometimes with many parts or spinning layers

motor: A device that converts a form of energy, like electricity or heat, into a type of mechanical movement, like spinning

NASA: An acronym for the National Aeronautics and Space Administration, an agency of the United States government that is responsible for space travel and research into space-related fields

neodymium: A chemical element in the periodic table used in creating very powerful magnets

nuclear reaction: An energy source created by splitting apart or adding to the nucleus of an atom

offset: Being away from the center or middle of an object

ohm: A unit used to measure electrical resistance

output: The result of a device's processing based on an input

parallel circuit: A type of electric circuit in which all the components receive the same voltage

Peltier junction: A two-sided device that can turn the difference in temperature between the two sides into an electric current, or an applied electric current that can make one side cold and the other side hot

permanent magnet: A material that can hold a magnetic field for a long time, such as refrigerator magnets and neodymium magnets

perpendicular: Being at a 90-degree, or right, angle from another surface or line

photon: Particle element of light

photoresistor: An electrical component with a resistance that changes depending on the amount of light hitting it

pneumatics: A technology that uses pressurized gas, or air, to move actuators

polarized: A state of a type of electrical component, such as batteries and LEDs, that allows for the passage of electric current only if the component is connected in a certain way

program: The act of telling a computer or microcontroller what to do, and how and/or when to do it, using a special language, or code

propeller: A type of fan that pushes air or water either forward or backward, depending on the direction in which it spins

prototype: The first model of a robot, machine, or structure that is used to perfect future models

pulley: A wheel with a groove that holds a rope, used to lift heavy objects, in which the rope is tied or connected to the object, passes up around the pulley wheel's groove, and can be pulled at an angle, decreasing the effort it takes to lift the object

radiation: Energy that is released in the form of waves or particles

Renaissance: The period of European history that marked the end of the Dark Ages and the beginning of science as we know it today

resistor: A device that slows the flow of electric current in a circuit

roboticist: A person who invents, designs, builds, programs, and fixes robots

robotics: A specific type of engineering that focuses on working with robots

satellite: A human-made device, or robotic probe, that orbits Earth in space

science: The process of trying to understand how something around us works by observing, or looking at it, guessing why it does what we observed, and experimenting to see if our guesses are right or wrong

semiconductor: A material that can conduct electricity under certain conditions

sensor: A device that can detect and send information about its surroundings to a controller

series circuit: A path for electric current in which all the components share the total voltage of a power source

servo: A type of gear motor that can sense its position, or where it is facing, in a set range of motion

solar panel: A device that converts the energy from photons into electric current

soldering iron: A tool used to melt a soft metal (called solder) to connect two harder metals (usually copper in electronics)

solenoid: A type of actuator that uses an electric current passing through a coil to create a strong magnetic field

Space Age: The period from when people first sent satellites into space, with Sputnik 1 in 1957, to the present

space probe: A robotic spacecraft sent out past Earth's orbit, to explore the solar system and beyond and send data back to Earth

STEAM: An acronym for science, technology, engineering, art, and mathematics

stimulus: A change in surroundings that can be detected by a sensor (plural: stimuli)

technology: The act of using scientific discoveries to create helpful tools, processes, or gadgets for people to use in their work or daily lives

terminal: The end point of a wire, battery, or electrical path on a device or in a circuit

thermometer: A device used to measure temperature

tilt sensor: A special type of switch shaped like a tube with a metal ball inside that can roll back and forth when it slopes to either side, allowing the ball to make a connection on the end of the tube, which then tells a controller which way a surface is angled

transistor: A type of semiconductor that can amplify, or increase, a signal and also act as an electronic switch, to either block or allow current to flow, such as is used by modern computers

troubleshooting: The process of testing different parts of a robot, circuit, or machine to figure out what is wrong if something doesn't work correctly

truss: Support beams in a structure, usually connected in a triangular shape

tutorial: An in-depth guide on how to learn a specific subject or do a specific project

ℱF (or mF): An abbreviation for "microfarad," a measurement of capacitance, or the storage of electrical energy

volt: The unit of measurement of electric potential

voltage: The difference in electric potential, or electric pressure, between any two points in a circuit

Index

About the Author

BOB KATOVICH realized his true passion of teaching electronics, engineering, and science while instructing kids of all ages at Robot City Workshop, a headquarters for hobbyists and tinkerers in Chicago. As Program Director, Bob organizes after-school programs and writes curriculum for kindergarten through college students. Recently, Bob has expanded his mentoring to Detroit elementary school students.

CPSIA information can be obtained
at www.ICGtesting.com
Printed in the USA
BVHW090745231020
591543BV00001B/1